THE FARM
AT BLACK HILLS

*To the men and women who have lived upon
and worked the land* — we look at photographs
of our ancestors with pride, but as they look
back at us are they equally proud or have we
failed them?

To my dear family and good friends — support
from loved ones is the fuel we need to keep
our engines running.

To Russell and May — who have been working
for over 64 years on the Forrester family farms.

Royalties from sales of this book
will go to Rural Women New Zealand

THE FARM AT BLACK HILLS

FARMING ALONE IN THE HILLS OF NORTH CANTERBURY

Beverley Forrester
with John McCrystal

RANDOM HOUSE
NEW ZEALAND

A RANDOM HOUSE BOOK published by
Random House New Zealand
18 Poland Road, Glenfield, Auckland, New Zealand

For more information about our titles go to
www.randomhouse.co.nz

A catalogue record for this book is available from the
National Library of New Zealand

Random House New Zealand is part of the Random House Group
New York   London   Sydney   Auckland   Delhi   Johannesburg

First published 2015

© 2015 Beverley Forrester

The moral rights of the author have been asserted

ISBN 978 1 77553 594 2
eISBN 978 1 77553 595 9

All photos private family collections except pages 18, 48, 72, 94,
122, 144, 168, 194, 220 and 236 © Guy Frederick and page 91 top
image Ref: 1/2-127240-F. Alexander Turnbull Library, Wellington,
New Zealand. Bottom right image by Johnston, C (Mr):
Photographs of the Bell, Featherston, Johnston and Robinson
families. Ref: PAColl-5564-002. Alexander Turnbull Library,
Wellington, New Zealand. Bottom left image courtesy of Waipara
County Historical Society, Private Photograph Collection.

Design: Carla Sy
Cover photographs: Guy Frederick

Printed by Griffin Press

FOREWORD

BEVERLEY FORRESTER, IN COMMON WITH her mother and her grandmother before her, is the type of member that every charitable-based community organisation would love to have and keep close.

Beverley has had a connection with Rural Women New Zealand for 40 years and has always been a loyal, innovative and active contributor. A true rural woman, she turns her hand to many tasks, from presenting a wool fashion show at The Beehive to baking for a neighbour in need.

Beverley's story is one of hard work with clear vision, talent undaunted by life's challenges, and a love of people, sheep and wool.

Her story is a good yarn. Travel the thread with her from rural North Canterbury to London, France and New York.

Liz Evans, JP, ONZM
National President 2011–13
Rural Women New Zealand

CONTENTS

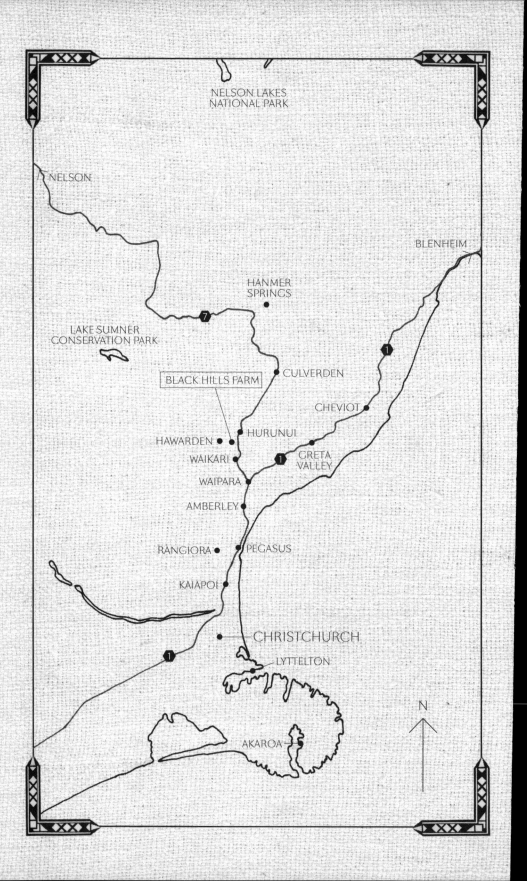

NELSON LAKES
NATIONAL PARK

NELSON

BLENHEIM

HANMER
SPRINGS

LAKE SUMNER
CONSERVATION PARK

7

CULVERDEN

BLACK HILLS FARM

CHEVIOT

HURUNUI

HAWARDEN

WAIKARI

GRETA
VALLEY

1

WAIPARA

AMBERLEY

RANGIORA

PEGASUS

KAIAPOI

CHRISTCHURCH

1

LYTTELTON

N

AKAROA

PROLOGUE

YOU KNOW, IT'S FUNNY how things work out.

You wouldn't believe that only a week ago I was stepping out into the hard, white glare of the spotlights on the catwalk at New Zealand Fashion Week up there in Auckland, 12 pretty young things — 10 girls, 2 boys — strutting beside me and wearing Beverley Riverina garments.

I line up a round of wood on the block and study the grain. I heft the axe and, after a couple of air-shots, I manage to split the round cleanly. I do it again. And again. When I have enough of it, I take an armload of firewood inside and stack it next to the wood-burner. That'll keep me going for the night.

In the cities — up there in Auckland, or in Wellington — if households go without power for more than a few days, you hear all about it. Well, it's been a week down here, and counting. There was a savage little southerly that knocked trees down all over the district. I lost a few. Russell has cleaned up a couple

Robert and Louisa Forrester

of them, but there's another one down in the road paddock that we'll get around to. The power and phone went out at the height of it. Telecom have worked out a system where my landline gets diverted to my cellphone so I'm not completely out of touch, although to get reception I have to climb the hill behind the house. You wouldn't think it, given I'm right here on State Highway 7, but I have no cellphone reception in the house, and no broadband. I suppose the power company will get around to fixing their wires eventually. Meanwhile, we'll just get on with it.

Luckily, the range is on gas, so I can still boil the kettle. I make myself a cuppa and take it out to the conservatory, where I sink into a comfy chair.

The daylight's going. A curious thing happens at this time of day, when the sun settles towards the mountaintops in the west. Over there to the east, the hills darken from the dun colour of tussock to black. They're still in full sunlight, but they turn quite black. It might be the low-growing matagouri scrub. It might be the shadows cast by tussock, or rocks protruding from the steep hillsides. There's one school of thought that says this is why they called them the Black Hills way back then. The name stuck.

I stretch my back and roll my shoulders. It's been a while since I've done so much wood-chopping, and it's been a bit of a shock to the system.

From where I sit, beyond the driveway and in the gap between the rounded hillsides, I can see the old homestead that the Perrott family lived in nearly 150 years ago. They wouldn't have been strangers to sore shoulders and the odd blister from the axe-handle, and nor would Robert and Louisa Forrester, who were living down there in the cottage when Sarah Agnes was born in 1892. By then, Robert and Louisa already had six children. Everything they did — from farming the farm to brewing a cup of tea — would have been hard work.

FROM HERE, I LOOK NORTHWEST, straight to where the hot, dry wind comes from more often than not. The backbone of the island rears above the plains in the direction of Hanmer Springs, 60-odd kilometres distant, and the peaks are lightly dusted with snow. It's a view you never get tired of staring at.

The willows down by the duck pond are beginning to get their new leaves, and the first blush of green is showing in my lucerne paddock down by the road. The paddocks on the plains beyond are green at the moment, too, but in a month the first tinge of brown will creep in, and from there it's just a question of how dry and brown it will get.

To my right are the Black Hills. The 2,000-acre property that was once named for them has passed to another generation, but I kept the Black Hills name for the block I'm farming, which used to be the Black Hills hogget block. My land — 250 acres — mostly lies behind me, stretching four kilometres back in the direction of Waikari (population: 800). Beyond the Black Hills, it's about 35 kilometres as the magpie flies out to Pegasus Bay, over the coastal hills. Waipara is 20 kilometres to the southeast through Weka Pass, and beyond it is Amberley and the main drag south to Christchurch. Hawarden is just over the ridge to the west. Sitting here, I'm pretty much right in the heart of what was once the largest sheep run in New Zealand, before it began to be broken up in the 1890s. I'm in the heart of the Hurunui District. I married into the district, into the story of Black Hills, and into the Forrester name. But my own family started out in New Zealand just down the road from here at Longbeach, near Ashburton, and moved up to Cheviot, so you could say I was closing the circle when I came to the Hurunui. Funny how it all works out.

I MIGHT NOT BE DOING it on the scale that George Henry Moore was doing it when he was running sheep on Glenmark Station a century and a half ago, but I'm running sheep on the same bit of land. Some of them are just outside, over the fence from the strip of lawn in front of the window. Most of the ewes have their lambs at their side. They're a kaleidoscope of earth tones — some black, some brown, some tan, and some a combination of all of them. It's funny how genetics works out and, when you're breeding coloured sheep, lambing time is a bit like Christmas — you never quite know what you're going to get.

This district, the Hurunui, is famous in the grand narrative of sheep genetics. James Little, a Scottish stockman, is credited with creating the Corriedale breed by the selective cross-breeding of Merinos and English longwool breeds down at Corriedale in North Otago from the 1870s onwards. After his employer died, Little moved to the Hurunui and settled close to Hawarden. It so happened that the Corriedale almost exactly suited the conditions we get here, and you'll occasionally hear the Hurunui described as the 'home of the Corriedale'. Most of the old farming families in the district have long histories with Corriedales, and the Forresters are no exception. We had a distinguished Corriedale stud, Heatherdale, from 1922 right up until 2000, when I decided I had to let it go.

While all those Forrester forebears probably turned in their graves when I relinquished the Heatherdale stud, I'd like to think they'd approve of what I've been doing since. I've been riding the resurgence of interest in wool that's arisen in the past decade or so, and that has taken me to some strange and wonderful places along the way. It's one of those ways that history knits one and purls one: in the early days, New Zealand sheep farmers were wool-growers, end of story. Then the frozen meat trade began, and the search was on for 'dual-purpose' sheep that would produce both a nice carcass and a decent fleece. The Corriedale fit the bill perfectly, although the same market forces saw the rise of the Romney in New Zealand, too. Funnily enough, the New Zealand Romney is quite different to the English Romney Marsh that it is descended from, and even noticeably different from the Romneys we had running around the hillsides in the early 1900s. The pressure to produce animals with a good fleece and good conformation has come slightly at the expense of fertility. But there you go: genetics is partly governed by the law of unintended consequences.

After the Second World War, and a brief boom when there were armies fighting in the cold in Korea, the world seemed to get it into its head that technology was the answer to everything, and that anything nature could do, man could do better. With that, we saw the rise of artificial fibres and a corresponding decline in the fortunes of wool. But in more recent times, people have been reminding themselves of the benefits of natural fibres, especially wool. There has been plenty of song and dance about Merino in the past 10 years, which has been marvellous for Merino farmers. And now we're seeing signs that mid-micron and strongwool growers might be seeing some action, too. Even at my end of the market, the niche end, there is lots of interest from around the world. Wool, and especially wool in its gorgeous natural colours, is making a comeback, and I'm pleased to be going along for the ride.

Still, 15 years ago, if you'd told me I'd have my own fashion label, outlets in the United Kingdom and the United States, and a financial interest in a woollen mill, I'd probably have laughed at you.

THERE'S THAT BLIMMIN RABBIT AGAIN.

From where I sit, I have a view of the little pond I have put in close to the corner of the house. A little half-grown rabbit has been popping out of the geraniums these past few days and hopping over to the pond. I'll have to fetch my shotgun and fix him, but I'll finish my cuppa first. That might give him time to get brave and come further out onto the lawn, which would be good, because I've got a plastic trough next to the pond, which has been leaking ever since the earthquake. I don't want to put a hole in the trough, too!

We're not too keen on rabbits in the Hurunui, or in the rest of the South Island, for that matter. It's hard for people who haven't seen it at first-hand to grasp what damage the little blighters can do. The Hurunui has been farmed by its share of hard men, but I've seen rabbits reduce them to tears. They've ruined people. We knocked them back with that virus a few years ago, but you'll never really be rid of them. They're just another one of the things that are sent to try farming folk, like floods and droughts and international commodity prices.

Of course, there are all kinds of ways you can turn a penny from a farm these days. I'm nearing the end of the long process of restoring the original limestone and cob farm buildings, built in 1862 (the same year as the Christchurch Anglican Cathedral), and stocking them with the kinds of implements the pioneers used. I have opened the place up to visitors, and they come by the bus-load to have a look at a working farm and to get a glimpse of what it must have been like for the old-timers. It's a pretty exotic experience for a lot of our overseas visitors. And you forget, when you live and breathe it all, that not many New Zealanders have access to farms these days, so even a lot of Kiwis who come and have a look around are part mystified and part in awe of it all. It's quite humbling to be reminded all over again of what a hard and lonely life our ancestors led.

Of course, for all the isolation of farms and farmers, none of us is an island. Community has always been a huge part of it all. Without the assistance, the goodwill, the empathy and even (occasionally) the concern of neighbours, none of us would have got very far. It was probably even more the case back in Robert and Louisa's day, when a trip to Hawarden was a big deal, let alone a visit to Christchurch. It's no wonder rural communities are so tight-knit: you're all in the same boat, after all. And that's why local events — A&P shows, the Ewe Fairs, the Calf Sales, the rugby matches, the Hurunui Races, even a Friday night at the Hurunui pub — are marked on your calendar in red ink. They're a chance to socialise, to enjoy someone's company other than your own, and to renew and refresh the ties that bind the community into a whole that is so much stronger and resilient than the sum of its parts.

THE SUN HAS DROPPED behind the ranges and the dusk has crept up the hillsides as I've been sitting and reflecting. I'm chuffed with the way my new conservatory keeps the heat, just quietly, but it's going to get cold later. Still no power. It's probably time to light the fire and a few candles.

It's also high time that someone told the story of this place. Of course, it's not one single story: it's a whole tangle of stories that are twisted around and through one another, like the Hurunui River with its many streams in its braided bed. You can't tell the story of Black Hills without telling the story of the Forresters. You can't tell my story without following it where it intertwines with Black Hills, the Forresters and the wider Hurunui. And of course, the stories of Black Hills, the Forresters and my own forebears, and the million stories of the Hurunui, are part of the much larger fabric of New Zealand's agricultural history.

So I suppose rather than try to tease it all apart, the best way is to tell it like it is, back and forth across it like the shuttle across a loom. That's what storytelling is, after all. It's like weaving, or spinning a yarn.

You'll have to bear with me a bit.

1
MILES AND PRICES

THE WEATHER'S WARMING UP, which is good news, because we're shearing the lambs today. Most Kiwi farmers won't do them this young, but if you want to use their wool in the traditional way, you need to remove the tip. The very tip of each strand is brittle, you see, and if it's left on it will break off in the carding process — the straightening of the wool fibres — and form little pills and clumps that will be left in the yarn. So we shear them early, at three months old, and then when we come to shear them a year on, they've got a lovely, long, strong staple with a nice square end. (Staple refers to the length of the wool fibre taken from the sheep. The wiggles in it are known as the 'crimp', and denote the fineness of the overall fleece — very fine fleeces have a zigzagged formation; coarser fleeces have a curvature.)

For this first clip, we don't shear them the way most Kiwi sheep farmers do, either, in a big shed with lots of stands at which brawny men and women in black singlets bend to their work with a radio blaring over the noise of the electric gear. We use the shed, or sometimes lay a tarpaulin out on the ground, but the whole process is just about noiseless — exactly as it was 150 years ago, before shearing machines were invented. The first shear is done by hand shears (blade shearing), an almost forgotten skill these days.

We also tend to shear our ewes just before lambing. It's amazing, really, but each strand of wool contains a full history of the animal's year. When your flock is happy and healthy and well fed and watered, the wool is strong. But if they're stressed, the wool reflects this by growing weaker. This patch on the strand is known as a break, and it can be a real pain when you're working the wool into yarn. So the idea is that you shear the ewes before the stress of lambing puts a break in their wool.

I WAS BORN BEVERLEY RIVERINA PRICE in Warkworth, just north of Auckland, on 31 July 1951, which will explain for people who don't know why I chose 'Beverley Riverina' as the label for my line of fashion woollens. Riverina is a name from my dad's side. It was my grandmother's name: she was born Riverina Marsom Pulsford Heales, a great niece of Richard Heales, the Premier of Victoria in the early 1800s, and the bloke after whom the township of Healesville, northeast of Melbourne, is named. Riverina Heales married a man named Alexander McLellan in 1902, but, although she had a son with him, she divorced him on the grounds of desertion six years later. Sometime after that — 1908, at best guess — she emigrated to New Zealand and married Samuel Thomas Pettigrew Price in Rotorua on 9 February 1914. Sam adopted

Samuel and Riverina Price

Clarrie, Riverina's boy, and Clarrie and the two children born later (Rivvie, in 1918, and Mansel, my dad, in 1922) regarded themselves as siblings.

We're not sure how or even when Sam got to New Zealand: he left England sometime after 1911, seems to have travelled to Australia, and then came on to New Zealand. Perhaps having signed on as a ship's carpenter. Funny, he lived with us for eight years when I was growing up and he never spoke about his past. I still have his carpentry kit, which he gave me when I turned 21.

Riverina Heales was born in Deniliquin, which is in the Riverina district of southwest New South Wales. This was, of course, one of the world's great sheep-farming districts: sheep farming, you might say, is in my blood as surely as it's in my name.

My mum's parents, Olive and Cyril, lived in Warkworth, but back then it was nothing like the close neighbour of Auckland it is today. To start with, it was cut off from Auckland by the Waitemata Harbour. There was a settlement of sorts on the North Shore, but it pretty much petered out just past Takapuna, and that was about the limit of civilisation, apart from far-flung farms. As late as the end of the 1920s, everything beyond Takapuna was known as 'the Roadless North'. Getting a motor vehicle as far as Warkworth was a popular adventure sport, with the only bit of easy-going being Orewa Beach at low tide. The rest was a mixture of slippery hills, sand traps and mud.

Reminiscing about those days, my grandfather, Cyril Phillips — Pop — recalled that he bought a car when he married my nan, Olive Miles, but for actually getting places he always rode a horse. For this reason, and because there wasn't much in the way of medical facilities at Warkworth at the time, if anything went wrong in the process of delivering a child, a mother was so far from help that she might as well have been on the moon.

A lot of New Zealand was like this, and the deaths of children or mothers, or both, in childbirth were regular occurrences.

As a precaution, Olive and Cyril Phillips travelled south to Grey Lynn and stayed with Pat and Charlie James, Cyril's sister and brother-in-law, in Wilton Road. Ona Louisa Phillips — Mum — was born there on 13 May 1926. She was just three days old when she arrived at Warkworth on 16 May 1926.

THE PHILLIPS SIDE OF THE FAMILY comes from Theddlethorpe, Lincolnshire, a village — actually a pair of villages — that are best known today for a huge natural gas terminal that is located there. Charles and Elizabeth Phillips and two of Charles's brothers, Joseph and Robert, arrived in Auckland from London in July 1859, aboard the *Whirlwind*. The voyage took 105 days — which was about average — but was notable both for the bad weather the vessel struck, and for the unusually poor quality of the food.

The *Whirlwind* arrived in Auckland pretty battered and bruised, and minus bits and pieces of her rigging, and a week or two after her arrival, one of her steerage passengers sued her captain (as the owner's representative) for breach of contract, on the grounds that the food he was served was well below the quality promised when he paid his fare. The salt beef was either off or was more bone than meat, he said; the biscuit was spoiled, and the rice heavily infested with weevils. His complaints were backed up by several other passengers, but the case was dismissed on a technical deficiency.

Soon after their arrival, the Phillips boys went to work clearing bush in the Dome Valley north of Warkworth. There were several sawmills in the area, turning the kauri forest into planks to supply the young colony's building boom. Warkworth was actually founded by a flour miller: its original name was 'Brown's Mill', after John Anderson Brown, but it soon became known as Warkworth after the village in Northumberland of that name.

Within a couple of years, Charles, Joseph and Robert Phillips had been joined by two younger brothers, Isaac and Horby, and soon the five Phillips boys were settled on their own land at a locality in the Dome Valley that was initially (and understandably) known as Phillipsville, but came in the fullness of time to be called Streamlands.

My maternal great-grandfather, Frederick Phillips, was born there on 3 September 1864 to Charles and Elizabeth. He inherited the farm from his dad, and in 1894 married Charlotte Lawrie, the daughter of a Scotsman, James Lawrie, and Elizabeth Johnston. Charlotte had been born in Ponsonby on 2 May 1874, a little after her grandfather, Hugh Johnston, was lost with his timber trading vessel en route to Sydney. Fred and Lottie farmed Streamlands until Fred's death in 1923, aged 58, whereupon the farm passed to Cyril.

Cyril was born in Warkworth on 16 February 1899, and married Caroline Olive Ellen Miles, known as Olive. They had one of the first cars in the district, and they proudly set off the day after their wedding to get photographed. This entailed driving to the Kaipara Flats railway station, loading the car, railing the car and themselves to Auckland, unloading, driving to the photographer's Queen Street studios, getting the photo taken, and then reversing the process. Thank heavens for the digital cameras on our phones, these days!

Top The photograph of Cyril and Olive Phillips taken in Auckland in 1925 after their wedding and an epic journey!

Bottom Cyril and Olive (Nan and Pop Phillips) after 65 years of marriage.

IT IS THROUGH OLIVE and the Miles side of things that my family was originally connected with North Canterbury. William Henry Miles, his wife Mary Jane, and their four children arrived in Lyttelton on the *Hereford* on 19 January 1878, after a fast and relatively uneventful run out from London of around 80 days. William was 25 and, like Mary (who was two years older than he), was from Perrygrove farm, near Coleford, in the Forest of Dean, Gloucestershire. He listed his occupation as an agricultural labourer; like most of the rest of the *Hereford*'s passengers, the family was immigrating with the assistance of the Canterbury Provincial Government. Under the assisted migration scheme, the provincial government typically paid half the fare, with the migrant (or their sponsor) picking up the other half. Usually they would pay a cash deposit to the provincial government agent in the United Kingdom, and the rest would take the form of promissory notes, to be redeemed as the immigrant found work and earned enough to pay down the debt.

Initially, the Miles family settled near Ashburton, at Longbeach, where William worked as a teamster for seven years. Then they moved up the road to their own land — another farm near Tinwald — where they lived for a little over six years. By now, the family had grown: William and Mary had nine children, four sons and five daughters.

In 1894, when the government subdivided Cheviot Hills Station and offered land for sale, both William and Mary Jane Miles were among the successful applicants: Mary, using her maiden name, Kear, was the first woman to own land in the district. The family travelled (as a newspaper obituary of Mary Jane Miles picturesquely put it) by covered wagon to their new home. On the title deed of his 50 acres (and 150 on lease-in-perpetuity), William is described as a 'contractor' — the farms in the Ashburton district were likely the 'few acres' that most settlers tended to own to keep themselves and their families.

In the Cheviot district — the name of the subdivision was officially the Caverhill Village settlement after the founder of the vast Cheviot Hills station, John Scott Caverhill — the Miles family ran mostly cattle. They suffered something of a setback in 1901, however, when a large earthquake caused significant damage to the Caverhill Village infrastructure, destroying the cream factory and breaking up most of the roading. The Miles family home was tipped from its foundations, and the coal range thrown onto the kitchen floor. It was fortunate that no one was injured.

Meanwhile, there was an upheaval of a different sort going on, as New Zealand had become embroiled in the first of the many foreign wars it would enter during the twentieth century, obliged — as the puffed-up rhetoric of the day had it — by 'the crimson tie of Empire' to help Britain out in the Boer War. One of those who volunteered to go was William (Bill) Howard Miles, the youngest of the four children whom William and Mary had brought out from England. Bill was living in the coal-mining town of Denniston at the time: his service record shows Private Miles's occupation to have been 'coal boy' (although according to family lore, he was working shoeing pit ponies under the Irish blacksmith, James Mateer), and notes that he was five foot eight inches tall, and had dark hair and hazel eyes.

Bill likely sailed in the *Gulf of Taranto* from Wellington on 6 April 1901, as he was part of the relief for the 6th and 7th Contingents. He must have found the action to his liking; rather than return home when the 9th Contingent arrived to relieve the 6th and 7th in April 1902, he re-enlisted, and was transferred to the 9th. He saw out the war and returned to New Zealand in July 1902.

Folded carefully into Bill's wallet all the while he was away fighting was a photograph of Alice, the eldest of the nine children of the blacksmith James Mateer, and his wife, Louisa.

Alice had been working as a housekeeper for none other than Richard Seddon in Kumara, before politics took Seddon north to Wellington in the mid-1890s. A little under two years after his return from the Transvaal, Bill made good on whatever promises he may have made before departing, and he and Alice Jane Mateer were married in the Church of England Mission Hall at Denniston on 9 June 1904.

In the same year, the Miles family and their in-laws, the Mateers, shifted to Matamata, where just a few years previously the grand dreams of Auckland miller and entrepreneur Josiah Clifton Firth had been strangled to death by his burgeoning debts. Firth had leased nearly 60,000 acres of ancestral lands from Ngati Haua — and had then been complicit in seeing the iwi's title to it undermined in the Native Land Court. His vision had been to grow wheat on the plains and rolling downs adjacent to the Waikato River. However, he was bankrupted, and his holding passed, through a series of sharp deals, to the Bank of New Zealand, and, when this institution suffered its own reversals at the end of the 1800s, the Crown's Asset Realisation Board assumed the title. Government policy at the time was to free up land for smallholdings, and Firth's block was carved up into 117 farms ranging from 47 to 984 acres, and included one run of 2,956 acres. The hope was that the new technology of refrigeration and the advances in artificial fertilisers would make small dairy farming viable.

It seems unlikely to have been mere coincidence that saw William and Mary Jane Miles and James and Louisa Mateer apply for blocks in the settlement at the same time as one another. William and Mary were successful in selecting a block of 590 acres adjacent to Springs Road, and the Mateers took up residence nearby. The Miles farmed their property for many years until William judged the time was right to retire, and he and Mary shifted into the brand-new township of Matamata. Mary died in 1936, and William two years later.

Of their nine children, only five were still alive by that time. Two of their daughters and two of their sons had died, Lance Corporal Stanley Miles being killed in action in Caterpillar Valley, Longueval, in France at the beginning of the Battle of the Somme on 15 September 1916.

Bill and Alice seem to have followed their parents to assist in the running of the properties. One of their daughters, Louisa Ivy (always known as Louie), was born in Westport in 1905; the other, Caroline Olive Ellen (my grandmother, always known as Olive, or Nan to me), was born in Matamata in 1907.

SAM AND RIVERINA PRICE, their son Mansel and their daughter Rivvie first arrived in Warkworth from Dargaville in 1925. Sam was to take up an appointment as the manager of the newly opened local branch of the farmer-owned stock and station agent North Auckland Farmers' Cooperative Ltd, which was founded in 1903. He and Riverina bought a house on the corner of Hexham and Lilburn streets in Warkworth.

Sam's territory extended 'from Kaiwaka to Mangawhai and then from coast to coast down to Auckland', as a newspaper report at the time of his retirement described it — essentially the area of the modern-day Rodney District. It was all but roadless at the time, and horses were more frequently used than cars. In the early years, Sam would spend anything up to 12 hours in the saddle getting from place to place, at a time when it took an hour to travel the 20 kilometres between Warkworth and Wellsford, and four hours from Warkworth to Maungaturoto, a distance of 50 kilometres and just 40 minutes' drive these days. In 1969, Sam told a *Lower North Weekly News* reporter that he pretty quickly came to know every strainer post in the county.

Even when he got a car, things were still pretty rough. The state of the roads in winter meant that you needed to fit snow chains to cope with the mud. Sam would routinely wear out four sets of chains each season, because there was no point in taking them off and fitting them again whenever he encountered a stretch of gravel, so rare were those stretches.

The job itself entailed being 'stock agent, fat stock buyer, merchandise salesman, farm adviser, monetary expert and to most a friend'. You'd have to suppose the last part of the role was, on occasion, the hardest: during the Depression, Sam had to watch as clients — mates — were squeezed off their land by a combination of debt and low prices. In the pit of it, he remembered, it wasn't worth leaving his desk to go and look over stock that had been offered for sale, as any commission he stood to earn would likely fail to cover the costs of running the car. When it was all over, Sam told his boss that if there were another slump he would quit. 'I couldn't have watched all that over again.'

The anguish all around him during the Slump must have been made even more acute for Sam by the fact that in September 1931 his wife Riverina had died after a number of years of poor health. This left him with two small children to look after, Mansel and Rivvie — Clarrie was now 27 — as well as working his very demanding job.

MANSEL THOMAS PRICE — my father — was just nine when his mum died. After that he was raised by a succession of housekeepers, the turnover doubtless having something to do with just what a handful the boy could be. According to a family story, one housekeeper announced she was quitting as Sam arrived home from work: with a look of horror etched on her face, she told him how she had found her little charge on the front path trying to saw the family dog in half! According to Dad, the dog had refused to move, and it seemed simpler to saw it in half to get past it. The young boy was reprimanded and the dog escaped serious harm. Mansel was enterprising as well as practical. According to another story, he worked out a little racket for himself where he would wag from school on sale day and remove a piece of harness from a horse in a neighbour's paddock, hide it, and then 'find' it again and receive a reward from its grateful owner.

By the time he was eight, Mansel was already an accomplished horseman and stock handler. In the Warkworth district, as elsewhere in New Zealand at the time, sheep and cattle were shifted about by droving. Mansel drove his first mob of sheep from Warkworth to Kaipara Flats when he was just eight, and regularly helped out with droving after that. By the time he was 12, he was regularly driving mobs from the Warkworth sale down the main highway to Puhoi. There is a photo of him in the Warkworth Museum when he was about 13, sitting on his white horse at the loading of a herd of cattle onto the Warkworth-built scow *Jane Gifford* for transportation across to Great Barrier Island.

In later life, he used to reminisce about some of his droving career's hairier moments. Stock from south of Warkworth were driven through town on their way to the markets, and it wasn't uncommon for them to wander as the mob strung out. On one particularly memorable occasion, a hulking Jersey bull gave Mansel the slip and charged up the steps into Bridge House, which then (as now) housed a rather refined sort of eating establishment. The cook shrieked and dived under a table, whereupon the beast did a quiet circuit, inspecting the fine chinaware and white tablecloths, and then left as it had come,

Loading the *Jane Gifford* at
Warkworth to take to Great
Barrier Island. Mansel is on the
white horse on the left.

by the front door. Dad also used to tell us stories about the local traffic cop, who used to insist that Mansel keep the cattle to the left-hand side of the main road going out of town, and to ensure that they went around and not over the traffic island in the middle of the roundabout. Mansel claimed he had trouble making the mobs understand such instructions!

Meanwhile, Mansel was attending school, first at Warkworth Primary School and then Warkworth District High School. He didn't harbour any great academic ambitions, though: he always used to say he aimed to be in the middle of the class, with as many behind as in front of him. Like a lot of country kids, he was just marking time before he could leave and get on with his life, which he did when he turned 14. He went to work initially in the grocery, hardware and drapery department of North Auckland Farmers (commonly known as the NAF), and then outdoors as a stock agent, as his father was the NAF stock auctioneer. He was a keen sportsman, playing hockey and tennis to representative level.

But it was war that really changed his life. Mansel was too young to enlist when war first broke out in August 1939, but he was keen, like able-bodied young men all over the district. In 1942, when he was still 19, he joined up with the Territorials. The Second World War came a lot closer to home when New Zealand declared war on Japan in December of that year. When a Japanese submarine was sighted off the New Zealand coast in March 1942, Mansel was posted along with his unit, the First North Auckland Infantry Batallion, to Okaihau near Kaikohe, presumably to hold back the Japanese Imperial Army if they chose to invade over the Hokianga bar. Sam, meanwhile, was playing his own part, acting as fat stock drafter and government meat supplier for the Rodney district, and serving on the rationing committees for fertiliser, benzine (as petrol was known in those days) and transportation.

In mid-1942, Mansel was inducted into the regular army at

'Overseas Camp' at Papakura, and trained as driver-operator in the Signals Corps. Mansel embarked on the mighty *Dominion Monarch* — a lovely cruise/cargo liner converted to a smelly, crowded troopship — for the European theatre with the 9th Reinforcements to the 2nd New Zealand Division on 14 May 1943.

After an uneventful crossing, spent in drilling and awaiting their arrival with a mixture of eagerness and trepidation, Mansel and his nearly 7,000 comrades reached Port Tewfik in Egypt on 10 June. They came under enemy bombardment as soon as they were docked, and Dad remembered the sky lighting up with the naval counter-bombardment mounted by the Allied warships lying at anchor in the bay, and the patter of shrapnel and spent ordnance on the *Dominion Monarch*'s deckplates. The adventure had begun in earnest!

By the time Dad settled into Maadi Camp, it was just about time to pack up again. The North African campaign was all but over, the 8th British Army — ably assisted by the New Zealanders under Bernard 'Tiny' Freyberg — having defeated Rommel the 'Desert Fox'.

Dad and the rest of the New Zealand Division were in the process of being redeployed to Italy. It was only as he was driving his truck down the ramp onto Italian soil that it occurred to Dad that he had no idea what side of the road they drove on in Italy. So he hedged his bets and stuck to the middle until it was plain that everyone else was driving on the right.

Compared with others of his generation, Dad was reasonably forthcoming about his war experiences, especially as he got older. He used to tell us proudly that as a non-smoker he was more comfortable than most, as he could trade hoarded cigarettes for little luxuries, such as getting his laundry done. Dad was involved in the entire Italian campaign, including the infamous battle for Monte Cassino, as far as the town of

Trieste, right up on the border with what used to be Yugoslavia (Slovenia these days).

But there, one day in late 1945, when he and some mates were diving into a swimming hole in a river, he misjudged the depth and broke his neck. So when he came down the gangplank in Auckland after being evacuated from Italy in September 1945, it was on a stretcher, with his prospects of walking, let alone resuming his former active way of life, very much up in the air.

ONA PHILLIPS — MY MUM — grew up on the same farm that her great-granddad Charles had carved out of the bush in the 1860s. One of her earliest memories was the time when, as a five-year-old, she disobeyed her father Cyril's instructions and, along with her younger sister Jean, climbed a large heap of slag (fertiliser) only to tumble off and sustain a nasty compound fracture of her right lower leg. Olive, her mum, swept her up and took her to Cyril, who rode off to fetch Dr Osborne. The good doctor's treatment was to splint the leg and immobilise it with two heavy sandbags — there was no plaster of paris in those days — ordering Ona to stay in bed for six weeks. How Olive managed to keep her there, Mum never knew! There is a picture of Jean and Ona after this episode, holding flowers up to the camera. Mum still has a bandage on her shin.

Mum remembers an idyllic childhood on the Kaipara Flats Road farm. Down the road, the Littins had a tennis court, which the local children were welcome to use. Another neighbour, Cyril Farman, built a nine-hole golf course on his land, and everyone made full use of that, too. The Bunkalls had a creek running through their property, with a swimming hole under the wooden one-way road bridge that spanned it. Mum remembered that if you were swimming when a car rumbled over the deck, shingle would sift through the planks and fall on you. It was also a great eeling spot, and the established protocol was for the older kids to order the younger ones into the water to splash about and scare away the eels before they would enter.

Ona went to school at Warkworth Primary, and each Sunday she, Jean and the Littin kids, Floss and Phyl, would make the eight-kilometre round trip on foot to Sunday School in Dome Valley. As she grew up, Ona became accomplished (as most country girls did in those days) at domestic crafts, such as knitting, crocheting, tatting, embroidery, quilting and patchwork, as well as the darning and mending that was needed to make clothes go the distance. She helped Olive with the routine domestic chores, and also to bottle fruit picked from their own trees or bought for the purpose. She enjoyed all of these tasks, but as she grew up she also took night classes in other practical subjects, such as copperwork, woodwork and pottery. Much to Olive's alarm, Ona was attracted to farming, and she helped Cyril grub thistles and move stock over her mother's protests that 'a woman's place is in the home' and 'farmwork is too heavy for women'.

Ona moved on to Warkworth District High School, where she did 'commercial' subjects — bookkeeping, typing, shorthand — but she left in early 1941 to enrol at Brains Commercial College, which was run out of the NZI building in Auckland's Queen Street. She boarded with her aunt, Cyril's sister Zöe, and Uncle Jack Mooney in Old Mill Road, Grey Lynn.

Brains Commercial College offered a one-year, intensive course in the same range of subjects Ona had been taking at Warkworth. Mum always said she had no idea how her parents managed to afford to put her through the course, as it was very expensive and there was no hint of a government subsidy. Ona passed her exams: by the end of the year, she could do shorthand at 150 words per minute and rattle the typewriter at 50 words per minute, which she modestly described as 'quite good'.

Ona's first job was for Captain (retired) Gillmore, a veteran of the First World War who had opened a shop in Queen's Arcade in Queen Street, supplying clothing and equipment to military personnel, especially the American officers who were beginning to appear in Auckland in great numbers after June 1942 and who had to purchase their own uniforms. Her role was running the office of the Officers' Equipment Depot Ltd — doing the accounts, banking, letter-writing, etc. It was classified as 'essential' to the war effort.

Outside work, Ona enjoyed a busy social time in the big city: the Women's War Service Auxiliary (WWSA) ran a servicemen's canteen in Queen's Arcade and Ona joined up. The WWSA held weekly drills in the Rutland Street Barracks, and every now and again a fleet of army lorries would be laid on to take auxiliary members out to Papakura Military Camp so they could be dance partners at camp socials — all under the steely gaze of their commanding officers, of course.

Fuel and food were strictly rationed, and luxuries of every description became impossibly scarce. If a consignment of silk stockings was rumoured to be about to arrive at Smith & Caughey's or George Court's, crowds of hundreds of young women would gather and jostle to be first in the door. Those who missed out would enlist a friend with a steady hand to draw lines down the back of their legs before they went out, to make it look as though they were wearing stockings. But the Americans could get nylons, so the shortage of silk stockings ceased to be such a problem — and the stocks of the Yanks rose.

Like most young women at the time, Mum was very impressed with the Americans. Most of them were in their early twenties, and already had a taste of the war in the Pacific. They were in New Zealand on 'Rest and Recreation' leave, and keen to have a good time. Mum recalled that their manners were impeccable, and they were very well paid compared with their Kiwi counterparts, so when a Yank showed up to take a girl to the movies (about all the entertainment that was on offer at the time), he commonly brought a 'shoulder spray', a bright posy of flowers, often orchids. These were out of reach of most New Zealanders, and you can probably trace the streak of anti-Americanism that runs through New Zealand society back to those days, when Kiwi soldiers, green with envy, would watch girls out on the town arm-in-arm with grinning Americans and mutter that there were only three things wrong with the Yanks: they were 'overpaid, overstayed and over here'.

Ona and her parents got to know a good few of them who stayed at the numerous camps in the Rodney district. Most of them returned to the Pacific War, and, of these, a terrible number died fighting.

Mum had lots of correspondents among the ranks of the American and Kiwi soldiers abroad, and she made some lasting friends: Ernie Grigg, an airman she met at a WWSA function, and Oswald 'Ossie' Horrobin, who survived the war but was lost with 28 others when the Union Steam Ship Company's *Kaitawa* sank off Cape Reinga in 1966.

IN EARLY 1945, AS THE WAR wound towards its close, Ona changed jobs and began working for the New Zealand Creditmen's Association in Wellesley Street, mostly as a shorthand typist. She found herself in a reasonably well-paid position, far better paid than anything she might have been able to hope for had the war never happened.

Ona was among the singing, dancing, weeping crowd that

wildly celebrated VJ Day (Victory over Japan) in Queen Street on 15 August 1945. Things got out of hand and, according to urban legend, over five tons of glass from broken bottles had to be swept off the streets the following day. But after the exultation, the hangover set in. The maimed and injured began to return from the battlefields onto the long road back to health and wellbeing — often in hospitals or in rehabilitation. The Returned Servicemen's Associations around the country kept tabs on where their members were sent, and organised for people from their hometown to visit them, to let them see a familiar face, perhaps, and to pass on news from home.

That's how Ona became reacquainted with Mansel Price, whom she knew by sight and by name even though he had been three or four years ahead of her at school in Warkworth. She saw his name on the visiting list supplied to her by the Warkworth RSA when she and another Warkworth girl, Peggy Macfarlane, went to the Military Rehabilitation Annex of Auckland Hospital one day in 1946. Mansel was in Evelyn Firth for rehabilitation and improving: it looked pretty certain that he would be able to work again, but his doctors told him he would never cope with rough ground and wouldn't ride a horse again. Sport and farming were off the menu.

Ona and Mansel met again a little later, when Mansel was best man at Stuart Smith's wedding to Floss Littin, and Ona's sister, Jean, was a bridesmaid. Since Jean had a boyfriend (Don Francis, her eventual husband), Mansel was paired up with her sister instead. Ona clearly made an impression on Mansel in the rehab ward and at the wedding, and, even though she told him primly that she was 'not interested in pursuing the friendship any further', he was (in her words) 'persistent'.

At this time, Ona was working and boarding in Auckland with her grandparents, Alice and Bill Miles, who had lately retired from farming Woodcocks Road in Warkworth, and now lived in Phyllis Street in Mount Albert. Whenever Mansel telephoned and asked for Ona, Grandmother Alice would flap her hand at Ona to get her to go outside, so that she could truthfully say 'She's not here.' Foiled by this strategy, once he was able to have day leave from hospital, Mansel would wait for Ona outside her office in Wellesley Street and take her home. It was a long, concerted campaign, but ultimately a successful one. 'I dodged him as much as possible,' Mum was fond of saying, 'but in the end I was caught!' Ona Phillips and Mansel Price — Mum and Dad — were married on 2 April 1949 in the Warkworth Anglican Church.

THEIR PLAN WAS TO GO FARMING. Dad was turned down for a rehab loan to buy a farm, because the medical advice remained that he wouldn't be capable of the physical work required. But, of course, that didn't stop him, and he bought a farm from Ernie Hall on Matakana Road in 1949. To call it a 'farm' is gilding the lily a bit: apart from a couple of rough paddocks, it was an unfenced block of gorse. Still, he reasoned, ever the optimist: if you can grow good gorse, you can grow good grass. Mansel resumed his job as a stock and station agent for North Auckland Farmers; in this respect, he was like most of the farmers in the district — part-timers who had a day-job as well, a lot of them in the cement works. Mum took a job as an office assistant at the Rodney Dairy Company, earning slightly more than Dad, and in the early years they worked the farm at night and in the weekends.

There was a fine, lone totara tree in one of the paddocks, with a caravan beneath it, and Ona's parents, Cyril and Olive, left the Phillips family farm to take up residence there shortly after Dad and Mum bought the property in 1949. They lived there until the mid-1950s, when they shifted to town, first to Lilburn Street, then to Victoria Street. Cyril died there in 1992, at the age of 93.

All of Dad's money had gone into buying his land, so there was nothing left over to spend on improvements. At the weekly, then fortnightly stock sales, he would look to pick up the odd in-calf heifer, which he would run for a few weeks before taking them back to the sales. He was pleased if he made a shilling on each animal. But in the winter he struck a deal with Trevor Wech, who grazed his four or five cattle on Dad's land in return for grubbing gorse and thistles.

The house at Matakana Road was a three-room cottage (a bedroom, a kitchen-dining room and a small lounge), whose interior walls were lavishly dented and punched in, because Ernie Hall had run wrestling classes there. Although there was electric lighting, all the cooking was done on a coal range, and food was stored in a safe — a cupboard with air-holes and mesh panels to allow the air to circulate while keeping out the flies, rodents and cockroaches.

The washing was done in the copper in the detached wash-house, which meant lighting a fire beneath it and stirring the clothes in the big copper kettle with a large stick. Afterwards, the clothes were wrung through a mangle before being pegged out to dry. Anyone who had kept a house in rural New Zealand in the previous hundred years would have been completely at home, although they would have been mystified (and delighted) by the electric lighting and — the sole luxury — the wireless, later commonly known as the radio.

Sometime between when I was born (on 31 July 1951) and when Noelene came along in June 1953, Dad built a second bedroom onto the house. Both of us girls were christened and later confirmed in the Warkworth Anglican Church. Around the time of Noelene's birth, Mum and Dad raised a mortgage and bought a second property, 57 acres of head-

41

high gorse on Woodcock's Road, directly opposite the land that Bill and Alice Miles had farmed between 1941 and 1945, and that Pop and Nan — Cyril and Olive Phillips — were still farming. The bridge over the river that runs through there was often called 'Miles Bridge', even though its official name was Cherry Bridge.

Dad pushed gorse with the tractor and bull-bars, and a mate of his, Ken Macey, bulldozed it into rows, and just about every night and every weekend, for what seemed like years, the whole family would go over and help as Dad fired the heaps. We would pick up the stray bits, and both Noelene and I got pretty used to the throb and redness around the millions of pin-pricks in our hands. But eventually the land could be put into crops. Ever since she was old enough to have her own say, Noelene has refused to eat swedes or turnips: she says she ate enough of them for a lifetime when she was a girl.

Still, it was progress. And things were definitely on the up when Dad came home one day in 1955 from the stock and sundry sale with a fridge and a washing machine aboard, both much pre-loved, of course. Eventually he picked up an electric stove, too, and 'we thought we were made', as Mum recalled many years later, 'and had everything we'd ever need'.

One day, a truck rolled up the drive with a dozen sheep aboard. Dad had undertaken to buy them, but hadn't had the money on him when he struck the deal. So the driver was expecting cash on delivery, but Dad wasn't home yet. Ona reluctantly dipped into her own money — the last remnant of her savings, as it turned out. When Mansel got home, she told him that, since she'd bought them, the mob was hers. He accepted this with good grace, and they seemed to find a way of making M & O Price Farm Partnership run smoothly enough. In fact, it ran for 50 years until their deaths, and Noelene and I run it still. Dad reckoned the secret was that Mum was boss of everything inside the garden fence; outside and on the farm was his territory. It must have worked, too, because they never really argued.

In 1960, Mum and Dad sold the Matakana Road property, paid off the mortgage on Woodcock's Road, and promptly raised another to buy a block from Fred and Ruby King further along the road. When he was asked by his work to transfer further north as a stock agent, Dad decided his family were too established in Warkworth and left his job with North Auckland Farmers. He knuckled down to work the two properties. He ran sheep and cattle, and there were usually a few ponies and horses about, too.

IN 1967, DAD'S HEALTH began to deteriorate. For a man who had been told he might not walk again, and then was assured he would never ride a horse or cope with walking over rough ground, he had done pretty well. He had farmed vigorously for the better part of 20 years, including mustering and droving (from horseback, of course), and he was a keen member of the local hunt and even took up polocrosse in his late fifties. He was certainly a physical man: he was determined to run a marathon before he was 70, and managed to complete a half-marathon on his sixty-first birthday; and he was a hazer at the annual Warkworth rodeo, which he helped to found in 1959. But pain began to limit his movements, and so he and Mum sold the Kings' property to the Milk Treatment Company and moved into a house that they built in 1968 on the original Woodcock's Road block.

They bought 10 Angus cows from Ted Wenzlick — Mansel's faith in his own eye for stock wasn't shaken even when one of the cattlebeasts died before he had signed the cheque — and started the Mahurangi Angus Stud at around the same time. Even when they had added 30 acres of Pop and Nan's land that adjoined their own, the Woodcock's Road property was soon too small to run a stud: there just wasn't the space to keep the boys and girls separate, so M & O Price Farm Partnership was obliged to lease land over the district, and that meant driving the beasts to and fro.

By now, Sam Price was living with them. He had retired from North Auckland Farmers in May 1954, after 26½ years' service, and had then moved to Long Bay to manage the estate of George Vaughan, who died suddenly in 1952 leaving a wife and four small children. He came to live with our family in 1965, and was there until his death on 15 November 1973, at the age of 84. He was buried in the Warkworth Anglican Cemetery, reunited at last with Riverina after 42 years apart. Because we lived so close to my grandparents, we also visited often. Whenever we felt like it, we'd saddle up our ponies and ride over the hill to spend the day with Pop and Nan. I was about 14 when Granddad Price — Sam — came to live with us. We didn't have a big house, so for eight years he was just part of the family, another face at the dining-room table. I still count myself fortunate to have been so close to that generation of my family.

Mansel sold his stud Angus herd to Stillwater in 1993 — and promptly bought a bunch of heifers back again when Stillwater announced it was reducing its numbers. He and Mum kept farming and breeding Angus beasts for years after that, and they were very active in the Northland Angus Association. That was another thing about Dad (and Mum, too, for that matter): service was a very important part of their lives.

Ever since the kindnesses shown him by the medical staff and the RSA upon his return, wounded, from war, Mansel was a very active member of the Returned Servicemen's Association, especially in their welfare work, reaching out to old soldiers in the community and looking after them. He was voluntary Welfare Officer for nearly 50 years, first for the Rodney RSA and then for South Rodney (which it became after the Wellsford branch split off), later renamed Warkworth and Districts. Mansel was appointed Branch President and then later Treasurer of South Rodney RSA. While he attended the committee meetings as Treasurer, Mum actually did the work: you had to be a returned serviceman to serve on the committee in those days.

By the time Dad resigned as Treasurer, Mum had been made an Honorary Life Member, the first woman to be so honoured by South Rodney, which entitled her to take over the Treasurer's role in name as well as in fact. This honour was conferred to recognise her work in supporting Mansel, and also in forming and acting as inaugural president of the Women's Section in 1961, which began with around 60 members. In this role, she also helped organise social functions for RSA members, and also made and embroidered a table cloth to be used on the chairman's table at committee meetings. It's still being used! She also assumed Mansel's position as Welfare Officer when he relinquished those duties in 2005, along with that of Pensions Officer. This was, of course, all voluntary work. By that time, Dad had been honoured many times by the RSA. He was awarded life membership in 1966, the 'M' badge in recognition of voluntary work and a certificate of merit in 1971. In the early 1990s, he was awarded the prestigious RSA Gold Star in recognition of service to the RSA and the community in general.

Dad also served on countless other organisations and committees — everything from the Rodney Pony Club, the Rodeo Club, the Warkworth Vet Club, the Northland Angus Cattlebreeders and the local branch of Federated Farmers to the A&P Show Society — several of which he was made a 'life member' of. He was a stalwart of the local Lions Club, which he joined in 1967. For his service in that organisation, for his RSA work and his work in the wider community, the Lions awarded him the Melvin Jones Fellowship, an international recognition of members who live the Lions' motto 'We Serve'.

In the 1990s, Mansel and Ona travelled first to the Lions International South Pacific Conference in Fiji, and then to the World Congress held in Chicago where a New Zealander was installed as World President. This gave them the taste for travel, I suppose. From the 1990s onward, Dad and Mum got about quite a bit, visiting many of the sites that Dad had last seen in wartime (an emotional journey for both of them), and also places of family significance. They had the thrill of tracking down Sam's family in Herefordshire, England. They also did farm tours in Australia, safaris in Africa, and holidays in the United States and Canada.

Meanwhile, they were farming. Then, as always, they were hospitable and generous in sharing the delights of rural life with outsiders: Mum and Dad had always been encouraging of schoolchildren, and particularly people with disabilities or special needs getting involved in farmwork. For a couple of years, they were also like parents to Yen Chiew, a Malaysian student who was attending Mahurangi College. Around 2000, they joined the Farm Holidays and Rural Holiday schemes, and hosted all manner of people — farmers, of course, but also schoolteachers and even USA High Court judges — on the farm at Woodcock's Road.

IN 2002, MUM AND DAD were approached by a property developer to see whether they would be willing to sell the town end of their land for development as a shopping centre. They had no real qualms about turning down this offer. But when they were approached again, this time by someone wanting to also build a retirement village, they were a little more receptive. Part of their first farm in Matakana Road had become the Totara Park Retirement Village; a portion of their last farm soon became the Summerset Falls Retirement Village. Both of them shed tears the day the bulldozers moved onto the property, but Dad, always one to look on the bright side, even now as his body and his mind began to let him down, took a keen interest in the work and was on very friendly terms with the workers and with the new owners. They named the approach road to the new complex 'Mansel Drive' in his honour. It is gratifying to see the road name appear on local maps and in addresses.

A little over a month before he was due to turn 87, Mansel suffered a combined heart attack and stroke and died some 30 hours later on 26 January 2009. He had been out on the farm only a day or two beforehand. He and Ona were just two months short of their sixtieth wedding anniversary. As Noelene told the big crowd at his funeral, he was fond of saying that you didn't get a second chance to make a first impression, and that you couldn't do better than give everything your best shot. I reckon my dad, Mansel Thomas Price, gave it his best shot, and it was plenty good enough. Dad himself had always said he was grateful to reach 'old age', as he hadn't been expected to survive his accident in Trieste, 64 years previously.

Dad left Mum on the land her family had farmed for three generations, but she wasn't alone. She had Noelene and me, and Noelene's husband and children, Lynley Alice and Christopher Mansel, but she also had her mum Olive, my dear nan, still going strong at 101.

2
FROM THE NORTH

SOME OF MY LAMBS are getting pretty hefty now. I've already done my first cull, and any that don't measure up have been sent away. It'll be time to move the rest down to Amberley to fatten up until just before they cut their teeth in July, which is when we'll sell them. (So long as they haven't cut teeth, they're still regarded as lambs and fetch a better price than they will if their teeth are through.) The way we've worked things out, we bring them to market at a pretty good time, when there's a shortage.

I keep the best of my ram lambs. Each year, I sell rams to the Agrodome up in Rotorua, and some of them find their way from there to all sorts of amazing places. I think the last lot ended up in Thailand, of all places, but they quite often go to Japan. Back in 2000 or thereabouts, I sold embryos and semen to Brazil. That was a oncer. I get lots of requests to send gene stock to the United States, but the hoops you have to jump through make it hardly worth the agony. And, of course, I sell a lot of rams locally, too. People often want to buy my show rams.

These days, I keep my stock numbers pretty low and run about 300 head. I could carry more, but I don't, for a number of reasons. For one, it's easier to do things biodynamically if your animals have got a bit of elbow room, I reckon. I think half the reason New Zealand sheep farmers have their work cut out with fly and worm control is that their flocks are all packed in tight. And it's less work if I'm not looking after all the stock that provide my wool, all the time.

A while ago I decided to take a leaf out of Te Mania Stud's book, after listening to one of their staff at the bull sales. What they'd been doing for a while at Te Mania, over there at Conway, was selling bulls to farmers and buying back the male progeny as steers — that is, bulls without testicles — fattening them on their own feedlot and selling them as their own genetic stock. Well, that made a heck of a lot of sense to me. So what I do now is sell rams to people who are keen to supply me with wool, and, provided I can be sure their husbandry is up to BlackHills' standard, I'm quite happy to buy the wool from those rams' progeny as authentic BlackHills gene stock. It's like a symbiotic relationship: the people who buy my rams have got a guaranteed market for their wool. I flatter myself that it's a bit of a drawcard for them to be involved in the BlackHills venture, too. And I get access to a whole lot more wool without having to do the hands-on work myself.

AT THE TIME I WAS BORN, my parents were still farming Totara Park on Matakana Road. My dad was a meticulous diarist — he kept a diary continuously for 65 years — and the entry for 31 July 1951 reads something like: 'Ona has girl. Fed out hay bales and collected 4 fowl eggs.' Contrary to the laconic way in which my arrival was greeted, I grew up on the farm surrounded by horses and the love and affection of my parents. My sister, Noelene, was born in June 1953.

Dad was still working as a stock agent in those days, and Noelene and I got to tour the district with him in his green Commer truck. He was a great singer, and knew lots of songs, mostly from wartime. He liked us to sing along with him, and quite often we'd get to our destination and he'd kill the motor and we'd sit there, singing, until the song was over and Noelene and I had mastered the words.

He also had us on the horses from the year dot. He used to take horses that had failed at the rodeo and break them in. Once he'd got them going, it was my job to ride them and get them used to the saddle. Lord knows how many times I got thrown! I expect that's why I have back problems today.

By the time I came to enter pony competitions, I was way ahead of most girls my age because I'd grown up riding. I was competing against the seniors while I was still a junior, first on a pony called Jimmy Brown, and then on a big Palomino called My Pal, a great horse, really, and one that Dad had broken in from the rodeo. My Pal was a bit of an escape artist, and we were always getting calls in the middle of the night to say he was off again. I can't remember whether he used to undo the gates or just jump the fences, but, either way, we had to head out in the middle of the night to find him. The pale wash of the headlights would pick him out going hell for leather up the road: he'd be white with foam, running along with his green canvas cover on. But he was a good competitor. I used to beat the seniors on him, which Dad thought was great!

Then we had a matched pair of blacks with pink diamonds on their noses: Dad's was Dinky Diamond, and mine was Twinkle. There was one fairly steep pitch on the Totara Park farm, down to a set of glow-worm caves at the bottom of a paddock, and I always used to fall off there. I'd hang on and hang on, but at a certain point — the same point every time — Twinkle would cough and over his head I'd go. 'One cough and she's off,' as Dad used to say. We found out much later that poor old Twinkle had tuberculosis, and there must have been something about the way the girth sat going down this little hill that set him off coughing.

Dad was pretty sought-after for his stock-handling skills, and I used to tag along with him as he went off around the district with his horse and his dogs and gave people a hand moving stock. He still did a fair bit of droving, even after he'd given away his job with North Auckland Farmers to concentrate on farming Kings', the place we had on Woodcock's Road. The remarkable thing about it — you had to keep reminding yourself — is that this was a man who was told he would never ride a horse again, and would never cope with the hard work that farming entailed. Dad's attitude in this, as in all things, is that where there's a will, there's a way. It might not be easy, and you might have to think outside the square to get something done, but everything was possible if you wanted to do it badly enough.

And often enough things weren't easy. Once, we were shifting stock for someone at Matakana when Dad's horse threw a shoe, so he got the owner's permission to leave the horse in a paddock so that he could come back later to re-shoe it. About a week later, he did. Soon afterwards, he got a phone call from a woman who asked him if he would come and shoe his daughter's horse.

'And before you say you're not up to it, with your neck and everything,' she said, 'I've seen you. I drove past you, and I saw you. You've done a couple of horses up this way lately, so don't say you can't do it.'

Dad agreed to do it.

'I couldn't tell her it was the same blimmin horse she saw me doing,' he shrugged, and rubbed his neck ruefully. 'I did the front legs one week and the back legs the next.'

DAD'S ATTITUDE TO LIFE — and it was one that Mum shared — was that if you decided to do something, you just got on and did it, without moaning, and gave it your best shot. It rubbed off on us, I suppose, because Noelene and I had a pretty good work ethic by the time we got to school. We did lots of sports, and then Brownies each week. When it came time to move from Brownies to Girl Guides, Mum and Dad gave me a choice, and I decided to do music instead; Olive and her mum, Alice Miles, were both pretty good on the violin and loved their music, and Uncle Clarrie was a wonderful pianist and violin player. I started learning the piano accordion, and I loved it — still do. The instrument I had was a great big, heavy thing. I can hardly lift it now, and I wonder how I managed it as a girl. I had to carry it for miles each lesson day to drop it off at the music teacher's place on the way to school.

Dad and I joined the Tauhoa polocrosse team, and participated in this and the Waitemata Hunt Club for years. You wouldn't think to look at me today that I was an athlete, but I did pretty well in my younger days. I was quite good at long-distance running — I've done three or four marathons and a half-dozen halves in my time — and back at school I showed some promise in the hurdles. Dad constructed a set of hurdles in a paddock and, never mind the lingering effects of his war injury, used to race me around this makeshift course. That was typical of the kind of support that both of my parents gave us. I did well in the school sports competition, and was selected to represent the Northland region. I came second there. Funnily enough, I ran into the winner of that event, Joyce Penny, just a couple of months ago, after 50 years.

My running must have inspired Dad, too, because he always said he'd like to have done a marathon by the time he turned 70. Well, in 1992, we had a couple of his relatives from England out staying with us, and Dad told them that he was going off to realise his dream of finishing a marathon on his seventieth. They would have done well to note his careful choice of the word 'finish'. They were awestruck when he returned home after lunch, slumped into a chair and announced he had done it. What he had, in fact, done was complete an 'accumulated' marathon. He'd been doing bits of the official 42-kilometre distance piecemeal over the preceding days, and on his birthday he knocked off the final 10 kilometres or so. It was still a great achievement!

LIKE BOTH MY PARENTS, I attended Warkworth Primary School. On my first day, I made friends with another Beverley — Beverley Hatfull — and we've been marvellous mates ever since. I did okay at school. About halfway through, they put me in a class with Evan Cross, a wonderful teacher, comprising about half a dozen kids from two or three of the other classes. It was called a 'performance class', or something like that: I don't for the life of me know what the selection criteria were — it wasn't as though I was particularly academically gifted or anything — but maybe it was characteristics like determination, or skills in problem-solving, or something like that. Some of my other lifelong friends were in the same set-up. Our class teacher was one of those people still memorable years later in your life.

I moved onto Mahurangi College after Warkworth Primary. It was pretty clear by then that I wasn't a high academic. I loved geography, as I recall — perhaps it was growing up interested in the places Dad had been during the war that kicked all that off — but I wasn't too good at English or anything like that. In fact, come Fifth Form and the first of the big external exams that you sat in those days, and I narrowly failed School Certificate. I went for a recount, but actually lost three marks. This was a big blow because Mum and Dad had decreed that the family wasn't allowed to have a television until after School C, in case it interfered with my schoolwork. Well, I felt I'd let everyone down.

I was a July baby, so I was sort of a year ahead of myself at school, anyway, so missing out on School C the first time around wasn't such a big deal. I repeated the Fifth Form and passed, second time lucky, and didn't have too much trouble with University Entrance the following year.

By now, I was working in my holidays and boarding at Te Atatu with Milton Miller, an Auckland chiropractor who treated Dad for his neck, and his wife, Beverley. Our families were quite close in those days, and have been ever since. The Millers were a city family, and liked coming to visit us on the farm, especially since they had a developmentally delayed toddler. Terence had suffered a complication soon after birth, and this slowed his development. He used to get about sliding on his bottom until he was six, but he loved coming to the farm, and it helped him to get a handle on numbers and language — he would count the chickens and the dogs and the horses, and get his mouth around the words. Terence is in his fifties now, and he still comes down and spends time on Black Hills Farm. His brother Brad, a chiropractor, lives in Los Angeles, and Brad's wife Julie is the USA and Canada distributor for BlackHills yarn.

I THINK IT WAS PROBABLY seeing the difference that Milton Miller's treatment made to the quality of the lives of his patients that got me interested in chiropractic. New Zealand didn't have a college of chiropractic. Back then the only real option seemed to be to travel to the United States to study, a five-year course at Palmer College in Davenport, Iowa. You had to be 22 to enrol. So I decided to have a look at occupational therapy instead.

Occupational therapy (OT) is based on the fact that you improve the ability of people with physical or psychiatric impairment and disability to perform everyday tasks, especially the ones that they want to do or need to do. An occupational therapist will look at everything, including environmental things, such as adapting the layout of a person's house or workplace with appliances to make ordinary tasks easier, or getting people to do purposeful activity such as arts and crafts to build their confidence in their own capacity to plan and execute a project. OT arose as a formal discipline in the early years of the twentieth century, but in the end it is an expression of the medical philosophy that you should put the needs of the patient first. That's what appealed to me about it.

The course was five years: three years' instruction, and then two years spent bonded to the New Zealand health system in return for the government subsidising the cost of tuition. I started on 22 January 1969. I boarded at Point Chevalier with a friend of Aunty Louie's, before moving out to the Millers in Te Atatu. For part of the year, I attended lectures at the OT School in the grounds of Oakley Hospital in Point Chevalier, and for the rest did practical placements at Auckland Hospital. From the outset, I had a sense that I had found my vocation.

At the end of my first full year, I, along with the rest of

my intake, was transferred to Christchurch. At first, we lived in the Christchurch Hospital Nurses' Home and travelled on one half day each week out to Sunnyside Psychiatric Hospital for psychiatry lectures. Then it was reversed, and we moved out to live at Sunnyside and commuted one half day a week to Christchurch for lectures. Like Oakley, Sunnyside was a pretty grim sort of place, and part of our work experience was to observe and even assist with the work the staff did with psychiatric patients. I saw electro-convulsive therapy (ECT) administered on several occasions: it was pretty horrible to watch, but in those days it was an accepted part of the treatment regimen. The wonder drugs that are used today were only just starting to come onto the scene back then, and ECT was the only effective tool at the disposal of psychiatric medicine. Margaret Bazley — Dame Margaret, these days — was the matron of Sunnyside while I was working there. What a wonderful woman she was.

It wasn't all work and no play in those days. The supervisor at the Nurses' Home would all but count us as we came in at the front door from our day's duties. Nine o'clock was curfew, and we'd all be signed in by then. But as soon as the front door was locked, we would head down the corridor and straight out the back door to where the motorbikes were parked, their riders — mostly young American men on their way back through Christchurch from Operation Deep Freeze, the American Antarctic programme — smoking cigarettes while they waited to whisk us off to dances and movies and parties and the like. I don't know whether our supervisor knew or suspected what was going on, but it's hard to imagine she didn't have some inkling.

In the third year of my studies, I was posted back to Auckland and then placed at Whangarei Hospital, working with children with congenital disabilities and with the ortho-paedic outpatients. After three months there and another three back at Oakley, I was placed at Porirua Hospital, working with those who were there for long-term rehabilitation. I'll never forget one day when we took a group of patients on the train to Wellington and then on to the zoo. Come home-time, we discovered that several of our charges were absent. They were all eventually located and mustered up, but, boy, did we get hauled over the coals for losing patients! At Porirua, I also did a stint working with alcoholics, amongst whom were some very prominent New Zealanders. You would be dying to tell friends and family who you had seen there for treatment, but, of course, patient confidentiality meant your lips were sealed. It was a $2000 fine and instant dismissal if you breathed so much as a word about your patients or their circumstances.

In January 1971, at the end of my three years' formal training, I was moved back to Christchurch to begin working out my bond period. My first placement was on Ward B1, the acute psychiatric unit at the Princess Margaret Hospital. We received attempted suicides, anorexics, people suffering from major anxiety disorders — all manner of diagnoses. The nice thing about my job was that your whole purpose was to treat the person as you found them, to do what you could to help them overcome the obstacles they were confronting, so that their lives could be made better. It wasn't always easy, but when it worked and you could see you were making a difference — it was a marvellous feeling.

For the first week after I started at the Princess Margaret, an office desk close to mine remained empty, with the name-tag of the new typist who was about to start sitting there next to her typewriter. I knew Lynley Tyson's name pretty well by the time she started. We got on like a house on fire right from the start, and about three weeks afterwards she asked if I'd like to come along to a meeting of the Avon City branch of Country Girls, a club for women like Lynley and me who lived in the city but had rural backgrounds. I became a regular at these meetings. And 40-odd years later, the same group still meets up!

In May 1972, the Princess Margaret's Ward K1 — the geriatric rehabilitation unit — opened and I began working there. People presented with a huge range of problems and disorders: everything from arthritis and orthopaedic trauma to the effect of strokes, multiple sclerosis and diseases such as Parkinson's. It was a terrific grounding for someone so early in their career. I had not yet turned 21, and I was loving my life and work.

LOOKING BACK OVER YOUR LIFE, you can usually identify a few people who really made a difference, whether by encouraging you at a crucial moment, or steering you in a certain direction, or both. One of those people for me was Joan Davidson, who was the Chief Occupational Therapist for the North Canterbury Hospital Board at the time. I'm sure I'm not the only person who has Joan to thank for life changes, as she was a marvellous woman. She was from Cheshire in England and was a truly formidable personality; a devout Christian, and a true and lifelong advocate for those with mental health issues and other disabilities. After she finished her role with the hospital board, she became a missionary, living and working in the slums of the Philippines; a couple of the other therapists and I supported her financially in a small way in this work. One of the last things she did before she died of cancer a couple of years ago was write a children's book that she hoped would entertain and inform children undergoing radiotherapy. That's the kind of person she was.

One Friday, I received a summons to Joan's office.

Oh, cripes. What have I done wrong? I wondered.

Apparently I hadn't done anything wrong. Joan sat me down and, without beating about the bush, said: 'They need someone to run OT at Templeton Hospital, and I want you to consider going and doing it.'

I went hot, and I went cold. I must have opened and closed my mouth a few times with nothing much coming out.

'Think about it over the weekend. But I want you to ring me on Monday and tell me you'll take the job,' Joan said.

Of course, I doubted I was up to it, but it was plain that Joan thought I was. It was an agonising weekend. It happens that way, doesn't it? Sometimes you get forced out of your comfort zone and it turns out for the best. But if you're not prepared to stick your neck out from time to time, you'll never really get anywhere. I decided I would take the job.

Joan didn't sound surprised when I told her on Monday morning.

'Good on you, Beverley,' she said.

So at the tender age of 22, newly graduated and not even finished with my two-year bond period, I started work as Charge Therapist at Templeton, with eight occupational therapy aides beneath me — all wonderful, as it turned out. Templeton Hospital and Training School was a 'psychopaedic' unit, and had opened in 1929 to treat (or, at least in the early days, house) children with intellectual disabilities. We had about 600, across a range of ages and with all manner of conditions and syndromes. They lived in villas, segregated by gender and by age. I'm quite sure it was working with these folk and wondering about the way that life deals you your hand of cards that got me started in my interest in genetics and heredity.

I'd had no experience bossing people around before — although Noelene might disagree. But my attitude was, and it has always been, that the best way to lead people is to work alongside them, and show them that you're not afraid to get stuck in and that there's nothing you'd ask them to do that you wouldn't do yourself. It seemed to work. Despite the doubts I had going into the job — what they call 'imposter syndrome' these days — I got on alright, and we had a good team there. We did some rather innovative stuff. For example, I was sent over to try to set up some courses for the first intake of inmates at Christchurch Women's Prison, which had been built practically next door to the hospital. I found that a bit too tough in the end.

We had other successes, though, like the Toy Library. One day when I was having a cup of tea with Lily McKelvie, a play therapist, she asked whether I could help her with something she wanted to do. Her idea was to buy a bunch of toys that could be loaned out to disabled students in the way that books are loaned out from a library. So that's what we did: we bounced ideas off each other for a fortnight or so, and I drew up a list of about 25 items that I thought could form the core of a lending collection of useful toys. We then established the first toy library in New Zealand at Christchurch East School, intended at first for disabled children, but soon catering to everyone. And just look at the toy library movement in New Zealand now!

There were some distressing moments in my early days, working with the disabled, the elderly and the mentally ill, but there were plenty of rewarding ones, too. And it strengthened my conviction that I was one of the lucky ones.

A few years ago, I was helping out in the Hawarden War Memorial Hall on one of the days when we have psychiatric patients come up from Christchurch for a day in the country. One of the patients suddenly said to me: 'I know you.' We got talking, and it turned out she had been a patient at Sunnyside when I was working there. As we chatted, it hit me that life had blessed me with a wealth of experiences and opportunities, while she had remained completely static, in institutional health care. It's the luck of the draw, I suppose, but it seems awfully hard.

I CARRIED ON AT TEMPLETON for a bit, commuting from my flat in Christchurch on a little Yamaha 50 step-thru scooter, rail, hail, snow or shine. At this time in my life, I met a man with whom I got on pretty well. When in 1973 he asked me to marry him, I was agreeable to the idea. Mum made me a hand-knitted lace 100 per cent wool wedding frock — just as she crocheted Noelene's, when her turn came. But then, after I'd had five good years at Templeton, I was asked to be a Christchurch-based tutor in Occupational Therapy, first for Petone Technical College and then for the Central Institute of Technology. I enjoyed this work, too, with its many challenges. Because the hospital was short-staffed, I sometimes had to cover for the OT clinical tutor in psychiatry.

Around this time, too, the Canterbury Occupational Therapists' Association asked me to represent them and New Zealand OT practitioners at the World Federation of Occupational Therapists Congress, to be held in Jerusalem in March 1978. They even raised $2,000 to help make it possible. This was a tremendous honour and, despite all the recent upheavals and the usual doubts and anxieties, I agreed to do it. And having decided to do it, I thought I might as well present a paper, so I set up a research project to furnish the material for it. Because it was envisaged that the paper would be published as part of the proceedings of the congress, I had to have it typed up in a special font. I'm no typist, I didn't own a typewriter, and nor could I find one anywhere in the area capable of producing the necessary font. In the end, asking around, we located one of those electric typewriters that had the right 'golf-ball': it belonged to the BNZ in Ashburton, and that's where my paper was typed up before being posted off to Israel.

It was my first trip outside New Zealand. I flew first to

Singapore, where I visited Gek Seng, a Colombo Plan student with whom I had trained in OT, and who was delighted to have me stay. I remember being struck by the strangeness of Singapore — little things, like the fact Gek's house had lino throughout, not a stitch of carpet, which I suppose is sensible in such a hot country. I also remember having a glass of water soon after I arrived and knowing right away that it was a mistake: I spent most of the 10-hour flight from Singapore to Tehran dashing down the aisle to the toilet.

I was hardly in top form when I arrived in the capital of Iran at 3am, and I wasn't exactly thrilled when I learned we were going to be delayed there. We were kept in the small, cramped transit area for 10 hours, but I wasn't about to argue. This was in the twilight of the Shah's regime, with political unrest and acts of violence becoming increasingly common. It was an act of terrorism that had the airport in lockdown: everywhere there were men in army uniforms, carrying guns, the first time I'd ever seen that. I was wearing sensible clothes — the advice had been 'not to dress like an American' — so I had on a long dress and a long-sleeved top, and I was glad of it. I remember sitting in transit writing letters to pass the time, surrounded by Arab men in their ankle-length white djellabas who were all watching, fascinated, as I wrote in a strange script from left to right, the opposite way to them.

It was a relief to be met in Jerusalem by people from the conference and taken to a comfortable hotel. After a shower and a rest, I felt a little closer to normal. My paper went well, and I enjoyed the conference and meeting people from all over the world. One of the local delegates was from a kibbutz, and they arranged a visit. It was Kibbutz Na'an — I use Na'an irrigation sprinklers here on Black Hills. I'd had no experience of boarding school or anything like that, so it was my first glimpse of communal living — so many bunks, and such enormous dining-room tables!

A real thrill was that afterwards I was picked up by a couple of Israeli girls who had been out to New Zealand on exchange. They had travelled three hours on a train from Tel Aviv just to see me. We saw the Wailing Wall and the Church of the Holy Sepulchre, and we took a noisy, hot, crowded bus through the shimmering heat and olive groves of Israel to Bethlehem itself. Seeing these places was strange: it felt a little bit like seeing a unicorn, or meeting Santa Claus.

It felt like I was spreading my wings. On my way back, I passed through Great Britain and did a short course on community occupational therapy at St Bartholomew's Hospital, right in the heart of London. Looking back, it was a big influence on me, and probably conditioned my whole belief on how a therapist could be of service to a rural community.

While I was staying in Los Angeles, I was contacted by alternative therapist Dr Milan Brych. We talked a bit, and Dr Brych must have liked the sound of me, because he offered me a job, travelling to people's homes when poor health or some other reason meant they were unable to attend his clinic, and setting up their homes for his therapies. It sounded like a good job and a wonderful opportunity, but it would be a big upheaval to shift. There would also be an awful lot of travelling around, and I wasn't sure I wanted that. And I had my work and family in New Zealand. After all that subsequently happened with Milan Brych, who was exposed as a bit of a charlatan, I'm probably lucky I made the decision I did.

In America, I bought one of those new-fangled microwave ovens. My grandmother, Olive, thought this thing was the most marvellous invention she'd ever seen. She'd seen live television footage of men walking on the moon, but that was nothing compared with the microwave: she couldn't believe that food could just cook itself like that.

AFTER MY RETURN, we moved to the outskirts of Ashburton, not too far away from where the Miles family settled at Longbeach. That's where I ran my first mob of coloured sheep, at first simply so I could use the natural-coloured wool in my own knitting. I bought 13 Romneys from the Oxford Paddy's Market, and began my experimentation with sheep genetics. I established a stud, which I named 'Bushlands Blacks', and registered in the Flock Book as New Zealand Flock #73. I used to take them to the local shows, which I loved doing. I also enrolled in a course in coloured wool production, reasoning that if I was going to get into this, I might as well do it properly. I also did courses at Telford and Lincoln that entitled me to do livestock and wool judging, which has been a lasting passion and opened all sorts of doors for me, down the years. Work was going well: I was asked to set the state finals Occupational Therapy exam a couple of years running. I had my sheep and my music — I was taking piano lessons to try to learn to use both hands, as it's different on the piano accordion. I was pretty happy.

And when the opportunity came up to shift to Ashburton to run their new occupational therapy unit, it looked as though all the pieces were falling into place. One of the highlights of this job was setting up an adult literacy programme. I got the summons one day to go to the office of Dr Chapman, the Medical Superintendent of the hospital. Just like all those years ago when I got the call to go and see Jean Davidson, my first thought was: What have I done now? But Dr Chapman handed me a letter and asked me to read it. It had come from the producers of a popular daytime TV show in Auckland called *Beauty and the Beast*, hosted by Selwyn Toogood. It was written by (or at least on behalf of) a bloke who lived in Ashburton, and who couldn't read and write but wanted to learn. The producers had sent the letter to the hospital, and Dr Chapman gave it to me.

'Can you set up an adult reading programme?' he asked.

I went off and investigated adult literacy — Mum was a big help, as she'd been doing a bit along these lines in Warkworth — and when I felt I knew what I wanted to do, I contacted the man who had written the letter and advertised in the local paper for others who might be interested. We pretty quickly had seven or eight applicants, all employed in good jobs around the Ashburton district, but all quite unable to read. We had our first meeting of prospective tutors at Jan McGregor's place. Cathy Hill attended — a teacher at Ashburton College who wanted to be involved.

In the end, we based our course around the adult education programme that the Hills knew at Ashburton College. As well as being involved in the foundation of the literacy programme, I had a couple of students myself, too: a girl who was working as one of two secretaries in a business in Ashburton, and whose work colleague, who had always acted as a human dictionary for her when she couldn't spell a word, was leaving; and the second in a political party, whose minutes of meetings were perfectly incomprehensible to everyone but himself. This bloke was keen on cars and engines and stuff, and the way I got him interested in reading and writing was to get books and magazines about cars, motivating him to work on his language skills. It was all about thinking about the problem laterally, as you do with occupational therapy. And, like occupational therapy, it was wonderfully rewarding work.

I WAS REALLY HAPPY, but it didn't last. My marriage ended after nine years. We both had our own ambitions and hopes, but gradually our friendship — which was the basis of it all — was placed under terrible strain. In the end, it was clear to us both that it had to be called off. So often in these situations, your in-laws stick to their own side of the fence. But mine didn't, and I am forever grateful that Alan and Bertha maintained their love and support of me right through this awful period, and never wavered in it to the end of their own earthly lives. They got on really well with my own parents, and, even after we parted, this relationship continued. Alan also had a bachelor brother, Ken — Uncle Ken to many of us — and he was also a tremendous enthusiast for anything I ever attempted. The wider family also remained friendly with me. A few years ago, one of my former sisters-in-law phoned.

'Beverley, I don't know anyone who lives on a farm anymore, and I want to show my grandkids what farm life is like.'

'Bring them up,' I said. 'By all means, bring them up.'

My former in-laws encouraged me to keep showing my sheep. The Leeston A&P Show was coming up, and I decided I would go. There, I was approached by a woman I didn't know from Adam, who shook my hand. 'I'm so pleased I've finally got to meet you. We're in the same stage of life,' she said meaningfully, 'and we both have a love of handcraft and coloured sheep.' She was a complete stranger in those days, but Helen McGregor (now Heddell) is now one of my closest friends: I was bridesmaid at her wedding, and she was later a bridesmaid at mine. She has also been a business mentor and a confidante. I'll never forget those words, the little gesture of solidarity, the enormous kindness behind it.

I DIDN'T FANCY STAYING IN ASHBURTON: it was too small a town, in the end, so I went back to Christchurch and got a job as second-in-command of the Christchurch Hospital occupational therapy department. Soon after that, I managed to sell the house in Ashburton, which enabled me to close that chapter for once and for all. Or so I thought. Of course, you don't get over an experience like the one I'd had just like that. I was terribly mistrustful of men; didn't want anything to do with them. It's probably fortunate that I worked in the kind of field that I did: being constantly called upon to nurture others brings out the best aspects of human nature, which were in danger of withering on the vine in my case.

I kept going to the A&P shows in the Christchurch area with my sheep, and had a bit of success here and there. As I began to build up a fair bit of respect amongst breeders of coloured and white sheep alike, I was soon getting called upon to do a bit of judging.

One day in March 1984, they asked me to go and judge at the Cheviot show. I knew roughly where Cheviot was, somewhere up there in North Canterbury, but I didn't know exactly. I was driving up — my uncle, Ray Viall, who worked in the Ford dealership in Papakura, had convinced me to buy a zippy little red Ford Laser Sports ('You need a nice car, Beverley. Not old. Reliable, because you're not mechanical, but something nice, something that makes you feel good about yourself. Leave it to me') — and I was going along on the main highway north near Belfast when a car shot past. It had two men in it, and swinging in the back window was the kind of tweed jacket that's practically the dress uniform for Canterbury farmers when they're off-farm. I was pretty sure they'd be heading to the same place I was. So I lowered my foot and stuck to their tail. If they changed lanes, I would change lanes. If they overtook someone, I'd overtake them, too. And when they parked in the paddock at the Cheviot showgrounds, I pulled in right beside them. They turned their surprised faces to me — it turned out to be a couple of blokes I knew well from around the shows. 'Oh, it's you, Beverley,' one of them said. 'We wondered who this lady was who was following us. We were getting our hopes up. What are you doing up this way?'

When I met my stewards — I knew them, too — I told them that I was feeling a bit let-down: I laughed and said the organisers had promised me that my stewards would be eligible blokes, and this pair were both married.

'Sorry to disappoint you,' one of my mates said. 'But don't give up — here comes one now.'

'One what?' I asked.

'An eligible bachelor,' he replied, pointing to a rawboned figure who was approaching. 'Over there. Jim Forrester.'

They introduced Jim to me.

'I know you,' I said. 'Last time I saw you was at your sister's wedding. You played the bagpipes.'

Jim grinned and blushed to the roots of his hair. Helen Forrester was a mate of mine from Avon City Country Girls. When her wedding reception was quite well advanced and Jim had likely put away a few whiskeys, he'd jumped up on one of the tables and winded his bagpipes. He'd struck up a tune and was pacing to and fro when the table suddenly broke in the middle. Down Jim went in a heap, and avalanches of cups and saucers converged on him from either end of the table. It's one way to get noticed, I suppose.

The Hawarden A&P show was supposed to happen a week after Cheviot, but it was postponed for a fortnight. I was asked up to judge again, and, along with my mates Helen and Peter Heddell, I was supposed to stay with David (always known as Doc) and Jan Sidey at Glenovis. There was no room at the inn, though, so they sent us to stay with Jan's cousin: none other than Jim Forrester.

The day after the Hawarden show, we were having lunch at the Hurunui Hotel with Doc and Jan and Jim, and I heard Jan saying, 'Ask Beverley. Bet you she'll go with you.'

'Ask me what?' I said.

'Nothing,' said Jim, going red again.

But a couple of days later I got a letter through the mail, very chivalrously asking whether I would agree to be James Kilsyth Forrester's partner at the dinner concluding the World Congress of the Hereford Breeders' Association, which was going to be held in Christchurch. That was the first time he asked me out, but not the last. Apparently, as soon as word had got around the district about me, there was pretty much a Hurunui-wide conspiracy to see me get set up with Jim; they'd been keen to get an occupational therapist living locally for years.

We were married on 14 June 1986 — a blimmin awful, snowy day — at the pretty little chapel at St Andrew's College, and soon afterwards Jim brought me home to the Black Hills, a bit of country that I would soon get to know like the back of my hand.

Snow on our heritage buildings, c. 2004.

3
GREENWOODS AND BLACK HILLS

LUCY, MY OLD SHEEPDOG, struggles to her feet, her ears pricked, staring intently out the glass door. I'm sitting with a cuppa in my conservatory, and her sudden movement startles me. She's an old girl, after all, and not much given to sudden movements. Then there's a crack, and the house sways and rattles a bit. It's the biggest shake we've had for a while, but not the only one. Since the big February shake in Christchurch in 2011, there have been millions of little rattles. You pretty much get used to them, but one on this scale gets you wondering where the centre was — surely not poor old Christchurch: people have only just reached the point where they can sleep again at night — and how bad.

The quake tapers off.

'That was a good one, girl. Wasn't it?'

Lucy looks at me with her bluish-white eyes, wags her tail

briefly and then settles down again. It's a peaceful winter's evening once more.

After a few phone calls, I find out that the epicentre of this one was near Seddon. People felt it down in Christchurch, but not too bad. Everyone around here seems to be fine.

My heart bleeds for the folk in Christchurch. I've still got friends with children living in their garage three years on from the February quake. We do what we can. Every now and then, a bus-load of people will come up to the farm, sponsored by the Lions — my friend Helen is one of the driving forces behind this — and these city people will climb off and do the tour and then often just lie in the paddock, where the blue sky, the quiet and the warm earth soothe their frayed nerves.

THE LAND THAT BLACK HILLS FARM occupies started out as seabed, way back before the ancient continent of Gondwana broke up and the world's landmasses drifted into their familiar configuration. Or so the geologists tell us. Up where the Hurunui rises, the bones of the country are visible, the Torlesse formation, grey and black rocks called greywacke and argillite — that's sandstone and mudstone to you and me.

These were formed at the bottom of the ancient ocean off the coast of Gondwana, where the sediments washed and settled. This area of seabed was uplifted by the collision of tectonic plates and formed part of the east coast of Gondwana. Eventually, a portion (which geologists call Zealandia) broke from Gondwana and decamped to our part of the globe, down under, crewed by the ancestors of the critters that have lived here ever since. As the Earth froze and thawed and froze again, sea levels rose and fell, and much of the base rock was intermittently — for periods of a few million years here and there — covered with shallow seas in which plankton and shellfish lived. These went the way of all flesh and their shells settled to the bottom, and this layer of chalky shells was turned over time into limestone. Shells and plankton weren't all that lived in the shallow, warm seas during this phase of things. The fossilised remains of ancestral fish, sharks and marine dinosaurs — truly horrible beasties, like mosasaurs and plesiosaurs — were collected from the Waipara riverbed in the 1850s and 1860s. The geologists reckon we probably had dinosaurs of all descriptions on the land, too, but their fossils haven't been preserved.

All the while, Zealandia was being crumpled up by the slow-motion collision of the tectonic plates it was straddling. The Pacific plate to the east slid beneath the Australian plate to the west, and the margin of the Australian plate buckled up. That's where we got the Southern Alps. And as the Earth entered its freeze cycle again and sea levels fell, the bits of Zealandia adjacent to the spine of the Alps emerged. That's where New Zealand came from.

We've had several more freeze-and-thaw cycles since the construction of our basic landforms, and the advance of glaciers up there in the mountains has ground the rocks into powder, which the melting has then carried down towards the sea, which has itself risen and fallen, in turn inundating the coast and receding again. The clutch of lakes near the headwaters of the Hurunui are evidence of the last big freeze a few hundred thousand years ago: they were the terminal lakes of a glacier that slumped down from the Dampier Range, bulldozing a whole lot of rubble before it. The rivers of the region — the Hurunui, and also the Waipara and the Waiau — have carved gorges through this layer-cake of different rocks, and deposited the silt and gravel in alluvial fans in the lowlands, where they sprawl in lazy braids; between floods, that is. Erosion has stripped away softer material and left the harder, deeper layers to weather here and there: the greywacke summits of some of our more prominent peaks, and limestone outcrops such as good old Frog Rock and The Deans, and the countless limestone scarps scattered across the district.

That's how the Hurunui district got made. Of course, Mother Nature never rests easy, and there's been a whole lot of tinkering since. The tectonic plates are still sliding. Away from the main boundary — the Alps — the land has been pleated and folded, and you can actually see these pleats and folds in the rolling hills and ridges of the district, especially where we are. Earthquakes are pretty common. Quite apart from the Alpine Fault, there are big, active faults up near Hanmer, and a web of hundreds, if not thousands, of smaller faults and folds between the mountains and the sea. They let go every now and then, and we have a bit of a shake, rattle and roll.

LOOKING AT THE HURUNUI DISTRICT these days, you'd be forgiven for thinking that it's always looked like it does — river flats and pasture land giving way to brown, tussocky hills, marching and mounting towards the rocky spine of the Alps. But funnily enough it's looked like this for only a few hundred years — less, if you're talking about the pasture with its English grasses and the blackberry and willows in the riverbeds. For some time since the Alps and their hinterland rose from the sea, they have been covered with vegetation: the remains of this, laid down in soupy swamps during the Earth's hot flushes, have been preserved as the coal measures that occur here and there throughout the district. Ice will have wiped the slate clean periodically, but after the peak of the last Ice Age, around 22,000 years ago, the forest staged a major comeback. The lowlands were blanketed with podocarp forest — totara, matai, kahikatea — giving way to beech near the tops. Birds, including moa, were all over the place.

The first human beings to come to the district were the Waitaha, an early wave of Polynesian settlement. They mostly stuck to the coast, exploiting the rich kaimoana resources of the rocks and the rivermouth, if the traces they have left — middens and ovens — are anything to go by. But they made forays into the forest on hunting expeditions, too: the ovens of ancient campsites have been found all the way up to the headwaters of the Hurunui, and just above Waikari there is a set of wind-sculpted caves that those first people used as shelters, doing a bit of doodling on the walls with charcoal from their fires in the evenings, or when they were pinned down like musterers by days of foul weather. They were hunters, gatherers and the district's first taggers, too!

But the graffiti they left wasn't their most dramatic legacy. Very soon after they first came to the area, the Waitaha set fire to the forests. We know this, because scientists drilling cores from the bottom of the lakes up at the head of the Hurunui can show you where the sediment layers bearing the pollen of the podocarp and beech forests suddenly give way to charcoal, and then to layers bearing the pollen of bracken and tussock. This matches the evidence from further south in Canterbury, and further again in Otago, where forests were also burned off. No one knows why just yet. It's been suggested that fires were set to flush moa out of the forests, and that these fires got out of hand. Or that the forest was deliberately burned to make travel easier, or to plant crops (although there is no evidence of large-scale attempts to grow anything in the South Island, which didn't suit Maori staples like kumara). We'll probably never really know what it was all in aid of.

The Waitaha were probably in occupation of this denuded

landscape for a few hundred years before they were conquered and absorbed first by Ngati Mamoe and then by Ngai Tahu encroaching from the North Island. Not much of the culture and traditions of the Waitaha was preserved: Ngai Tahu don't have any stories about how the rock drawings in Waikari and other parts of Canterbury came to be there, for example, and they don't have anything much to say about the fires, either.

Like the Waitaha before them, the more recent iwi occupations were based on important coastal sites like Kaiapoi to the south and Kaikoura to the north, but the Hurunui district became significant, too. The Maori soon found the saddle between the headwaters of the Hurunui on the eastern side of the Main Divide and the Taramakau on the western side, and this became one of the major routes by which they gained access to the West Coast, with its undisturbed forests teeming with birds, and the cold, swift rivers to the south where pounamu — greenstone — was found.

THE FIRST EUROPEANS to settle the South Island stuck to the coast, too. There were shore whaling stations at Kaikoura and on Banks Peninsula, and even one for a while on Motunau Island just out there in Pegasus Bay. As early as 1839, a Sydney-based whaler, William Barnard Rhodes, released cattle and sheep at Akaroa. It wasn't really an organised farming attempt: a couple of years later, a ship was equipped with stock, agricultural equipment and personnel to establish a farm on a block of land that Ngai Tahu were willing to lease, but the ship disappeared without trace, as ships quite commonly did in those days.

In 1843, the Deans brothers, John and William, who had

arrived in the first bunch of settlers sent out by the New Zealand Company to Wellington, made their way south in search of the nice, flat farmland that they had been promised and that Wellington had failed to offer. They established themselves at a site Maori called Putaringamotu and that the Deans boys renamed Riccarton, after the Scottish town from which they hailed. This was the first permanent European settlement on the Canterbury Plains, and the land was leased from Ngai Tahu with the permission of the colonial government.

In 1848, the New Zealand Company hatched plans to launch a systematic attempt to settle the region with decent, god-fearing Anglican folk, and on 12 June of that year a deed of settlement was negotiated on the Canterbury Association's behalf by the colonial government's agent Thomas Kemp and 16 Ngai Tahu chiefs. Under the terms of this agreement, the whole middle of the Middle Island (as the South Island was called), from Otago north to the Hurunui and right across to the West Coast, became the property of the Canterbury Association. The sale was conditional on reserves being set aside to keep sacred sites and important food resources in Ngai Tahu hands, but, in a mixture of ineptitude and bad faith, only a fraction of what should have been reserved was actually handed over to Ngai Tahu.

The first settlers under the Canterbury Association scheme arrived in 1850 aboard four ships, the *Cressy*, the *Charlotte Jane*, the *Randolph* and the *Sir George Seymour*. Four more ships followed close behind. Most of the people they brought out were tradespeople who were supposed to become the citizens of Christchurch, the town that had been pegged out by Canterbury Association surveyors on the flat, swampy ground over the hills from the port at Lyttelton. But the rest had ambitions to be landowners, and the carving up of the Canterbury region into runs began. A rental of £3 per acre was payable to the Canterbury Land Office where a licence to occupy was granted. Little wonder that people began looking further afield: anyone who had travelled between Port Nicholson (Wellington) and Port Cooper (Lyttelton) would have done so in a small vessel hugging the coast, and would have noticed the promising-looking land that lay just north of the boundary of the Kemp Purchase. It wasn't long before the scramble was on.

THE FIRST PAKEHA TO SETTLE the Hurunui district were the Greenwood brothers. Like the Deans brothers, Yorkshireman James Dent Greenwood and his two younger brothers, Joseph Hugh and Edward, were among the original New Zealand Company colonists in Wellington: they took up land at Lowry Bay in 1841. Like the Deans and a handful of others — whom they knew personally — they quickly grew discontented with their prospects in the North Island, and perhaps with the assistance of W B Rhodes (with whom they were closely associated, having grown up only 30 miles from the Rhodes family on the Yorkshire moors), they set themselves up on a block at Purau at the base of Banks Peninsula. They had no legal right to this land, but eventually negotiated a lease with Ngai Tahu, the yearly rental for which was seven blankets and some printed calico, to the value of roughly £4.

The Greenwoods traded their produce up to Wellington in an open whaleboat, and one of their favourite spots for putting into the coast when the weather cut up rough was the relatively sheltered mouth of the Motunau River, about 15 kilometres south of the Hurunui mouth. They may have intended the stockyard they built there as nothing more than a holding pen for animals they were carrying, and the nearby hut as a kind of transit facility, but the attractions of the surrounding land didn't escape them, particularly as things at Purau began to turn sour.

In 1846, the Greenwood boys were held up by a notorious criminal named 'Bluecap' and his gang, a trio of 'ticket-of-leave' men from Hobart — prisoners who had served their sentence. James and Joseph were tied up, and Edward was forced at gunpoint to strip their nice, new homestead of all valuables. The bushrangers escaped by boat, making Edward promise to give them 20 minutes' start, a promise that he kept. Once they freed themselves, the Greenwoods alerted the local whalers to the fact that Bluecap and Co. were about to attack the Deans at Riccarton, and that attempt was thwarted. The criminals were eventually apprehended in Otago and sentenced to 15 years' transportation. This was no consolation for the Greenwoods, though, who had lost much of their wealth and all of their sense of security in their isolated bay. And in the same year, their rent was doubled by their Maori landlords. The Greenwoods made plans to relocate to Motunau. They first formalised and then sold their lease on Purau to William Rhodes and his brother George, and set about shifting their stock overland to the Hurunui.

Not that this was easy! They needed to cross 100 kilometres of largely untamed land and treacherous rivers on foot, without the benefit even of pack animals. James Greenwood set out on 9 September 1847 with a man named Edward Fisher, droving their half-wild cattle over Gebbies Pass to the scrub-covered Port Cooper Plains, as they were then known. They camped, but woke to find they were alone: the beasts had gone bush. One of the hands from Purau, William Prebble, joined them, and they managed to relocate their animals. They drove them on to the Waimakariri River (Greenwood spelt it 'Wye McReedie' in his diary) and, with difficulty, across. By splitting the night into two watches, they managed to keep the herd more or less together. William Prebble's brother Edward caught up with them on 11 September, and they got as far as the 'Rakahooui' (the Ashley) before camping. Edward Prebble fell asleep on his watch, and when everyone woke the cattle were gone again. After a day spent searching, they found them again in a swamp, in which several were stuck. They camped, freed the bogged animals, and made it on to the Double Corner (Waipara) River on the 13th. After another tough crossing, they found it rough-going through the rocky and scrubby ground inland from Pegasus Bay to the Motunau River, but, while they were congratulating themselves on a job well done, they had still that final river crossing to contend with. This took most of the 15th, but everyone — man and beast — arrived safely at the yards by 3pm. Just half an hour later, one of their herd, the aptly named Traveller, dropped a calf.

The Greenwoods farmed on 60,000 acres at Motunau — mostly cattle at first, but with a few sheep. As an historian of early Canterbury runs, L G D Acland, records, 'the chief station work in those days consisted of milking, keeping boundary on the sheep, and hunting for lost cattle'.

> The sheep were kept on the flat and the front of the Limestone Range, and there was everlasting trouble with wild dogs. [The Greenwoods] milked all the cows, making 50 to 80 lb of butter a week. The dry cattle were run further out than the sheep, but had to be looked over every two or three days, and even then often strayed as far as Waikari and Kaiapoi. When horses were brought up, they went back several times to the Deans's at Riccarton.

Like others moving stock through the district, the Greenwoods had a lot of trouble with the poisonous shrub tutu, which grew prolifically in the riverbeds. And for decades parts of Motunau were famous for their wild pig populations. Wild pigs damage pasture and are a threat to lambs and native birds.

Their produce outwards and station supplies inwards were ferried, as before, by whaleboat between Motunau and Port Levy on Banks Peninsula. On one of these runs, Joseph Greenwood's luck ran out. He left Port Levy on 14 October 1848 with two crew — a pakeha named Johnny Moles and a Maori man whose name isn't recorded — and with another of the station's boats under orders to follow them around three weeks later. When the second boat reached Motunau, they found that Greenwood's boat hadn't arrived. Coastal shipmasters reported seeing bits of a whaleboat washed up on the beach south of Kaikoura, and there seemed little reason to doubt that Joseph Greenwood and his mates had been drowned.

Just a year later, James Dent Greenwood travelled to Sydney on business. He was carrying a large sum of money, and must have been a bit careless in flashing it about. He disappeared from his hotel, and is presumed to have been robbed and murdered and his body disposed of. Edward, the sole remaining Greenwood boy on the block, had naturally had enough of New Zealand, and returned to England, having arranged to let the property and stock to John Scott Caverhill. It wasn't quite the end of the Greenwood family's association with the Hurunui, though. Another brother, George, came out in 1867 and took up Teviotdale Station. The Greenwood family farmed there until the death, a couple of years back, of Daphne Greenwood.

⇒——→

JOHN CAVERHILL, who was also known by his nicknames 'Darby' and 'White-haired Bob' (on account of his shock of white-blond hair), was born in Jedburgh, Scotland, only a little to the north of the English border, in 1821. He followed his older brother Tom out to Australia in the late 1830s, and worked for a while as a stockman employed by George (later Sir George) Grey. A little after Tom crossed to New Zealand at Grey's invitation, John followed; in 1848, Tom was drowned in a river in the Wairarapa while in Grey's employ, and John soon found his way to Port Cooper, where he went to work briefly for the Deans and then for the Greenwoods.

When Edward Greenwood upped sticks for England in January 1850, he struck a deal with Caverhill, whereby Caverhill would take a third of the profits from the Motunau wool clip, and all of the natural increase of the sheep and cattle. This seems to have provided the right incentive: sheep numbers increased rapidly at Motunau under John Caverhill's stewardship. Edward Greenwood had left him 981 sheep, 474 lambs, 141 cattle, 40 pigs, 3 horses and a mule in 1850. Four years later, Caverhill had 4,184 sheep; 8,256 in 1857; and 12,000 in 1858. Caverhill was noted for his longhorn cattle stud, and for his line of thoroughbred horses. He was successful in agricultural shows, and much sought-after as a judge of livestock.

Besides being a first-rate stockman, Caverhill was known as a supremely generous man, and a bit of a prankster. When Captain Joseph Thomas, of the Canterbury Association, and Caverhill's close friend Robert Waitt arrived on an exploratory ride to the Hurunui district, Caverhill invited them to cross a paddock. They noticed he was hanging back — at the same time they noticed a stroppy-looking bull eyeing them up from the far side!

Caverhill was also a restless spirit, it seems, and loved to explore. He travelled all over what was to become known as the Hurunui district, and north as far as Kaikoura as well. He liked the look of the land to the north of Motunau, and began grazing cattle there. He called the block Cheviot Hills, and the homestead he built there, The Retreat.

Meanwhile, others had arrived in the district to take up farms, too. A few of them were former British Imperial Army officers who had distinguished themselves in action in Crimea and in the Indian Mutiny. Some of the others were, like Caverhill, Scotsmen. This may have had something to do with the celebrated national canniness of the Scots: it cost only 10 shillings to lease land north of the boundary of the Canterbury block, compared with £3 south of it. Or it may have had to do with the determination of the Canterbury Association to keep Canterbury Anglican: as Presbyterians, even the well-established Deans came under heavy pressure to relocate.

By the time Caverhill was established at Motunau, there were several others in the area. Inland to the west, Charles Sidey and Robert Mason farmed Heathstock and Horsley Down. Inland to the south, Robert Waitt had Teviotdale (which Caverhill helped him to select). Frederick Weld and Charles Clifford had marked out a property adjoining Motunau on the coast to the north; they called it Stoneyhurst, although they didn't stock it until a few years later. Mount Deans Station occupied the land around The Deans and Weka Pass: in 1851, its new owner, Dr Samuel Hodgkinson, changed its name to Birch Hollow, although everyone else called it Doctor's Hills. Neighbouring this property, there was a station straddling the Waipara called Double Corner — an old name for the river itself, which bends sharply back on itself in a couple of places in its lower reaches.

AS THE REGIONS TO THE north of the boundary of the Kemp Purchase were settled, the search was on for a pass to the more established colonies at Nelson and Marlborough for the purposes of bringing sheep through, as stock driven overland from New Zealand vendors was going to be much cheaper than animals imported from Australia. At the forefront of this effort was Frederick Weld, who had already established farming interests in the Wairarapa in the North Island, and in Marlborough when he and his partner, Charles Clifford, applied for the right to occupy 58,600 acres of land to the south of the Hurunui rivermouth in December 1850. Permission was granted, and the new property was named Stoneyhurst.

Weld's first attempt to bring sheep overland from Marlborough in 1851 ended in failure, because his drovers abandoned the mob within two days' walk of Stoneyhurst. They had, they said, 'run out of tucker'. Few of the sheep were recovered. In March the following year, another mob of 1,500 ewes was sent by the same route — down the Clarence — in the care of Charles Clifford's brother Alphonso. Only one was lost en route, and two other mobs followed close behind.

With the 'Canterbury track' established, work on the infrastructure began, with regularly spaced accommodation (and watering holes) along with holding paddocks and sheep dips installed. There was a chain of these, from Top House, Rainbow, Tarndale and Acheron. Huge numbers of sheep were driven via this route — 24,000 in the 1857/58 season alone. It was difficult work, as one contemporary account records:

> Three hands and myself left the
> coast to take delivery of a flock of
> 1,000 sheep not far from Nelson on

the northernmost boundary of the Province. The sheep, having been yarded during the necessary drafting and branding for three days and nights without feed, by this time half-starved and twice as wild, strung away at a rapid pace ravenously seeking for feed. As a rule, to drive a mob of sheep for any distance takes not less than four hands with three horses saddled and one packed with provisions and baggage; a pair of blankets and a tin pot quart per man, a frying pan, tether rope, butcher's knife, bag of flour and a canvas bag of tea and sugar, a nob of salt and a few boxes of matches with a reserve of flint and tinder, in most cases makes up the list. During the day the sheep have to be got somehow over the deep sides of the hills and through long stretches of plains, as well as the tangled growth of the wild country, the rugged nature of the ground, and the greater part of the heat of the day will permit. It requires incessant care to prevent stragglers breaking away from the head and sides of the mob, and the rear of the flock made up of weaker sheep requires an equally constant look-out. The sheep were so wild and restless that it was necessary to keep a double watch, two of us took charge, the other two camping . . .

As soon as the first light of dawn began to pierce the blackness of the night, it was the signal to rise and start moving. We travelled till 10 am when the sheep would camp down. Then we got going again in the cool of the afternoon and would camp the sheep again at dusk . . . Today when we cross rivers on well-built bridges, it is hard to realise the difficulties of sheep drovers when they had to cross a river with a mob of sheep. Sometimes sheep take to the water quite easily . . . at others it seems as if all the king's dogs could not induce them to swim.

IN 1851, SQUATTERS WERE OBLIGED to have their boundaries formalised by applying for licences from the government. Many of the areas subject to applications overlapped, and some of those claiming land had no history of occupation whatsoever. A Wellingtonian named William Lyon rather optimistically applied for a block that included the area known as Motunau Black Hills (or sometimes Motunau Burnt Hut Hills: one old settler has been reported as saying that the name 'Black Hills' was given to them when a big fire left the tussock blackened for years after it swept through), inland from the coast. Another, named George Cooper Pawsey, applied for an area that straddled the boundary between Motunau and Teviotdale. Caverhill himself applied for a licence over the

Cheviot Hills area. Who knows what kinds of processes were followed in determining the merits of these claims. Lyon's application for Black Hills was denied, but Pawsey's claim to 'Bob's Flat' — named after the Motunau boundary keeper, who was only ever known as Bob — was allowed.

Even then, the granting of a squatter's licence didn't protect your property in perpetuity. If someone alleged that you were farming more land than your licence stipulated, they could request (and pay for) a survey to be done, and if they were proved right, they could be issued a licence to lease the surplus. Or you could buy the freehold to someone else's leasehold land. If you were cunning and strategic about how you set about doing this, you could effectively snooker a leaseholder by leaving him only scraps of land separated from one another by your strips of freehold, a process known as 'grid-ironing'.

And, of course, money talked. When South Australia run-holder (and future New Zealand Legislative Councillor) William Robinson decided that he would relocate to New Zealand, he appointed an agent to scout likely-looking properties for him. John Oakson, the agent, identified Cheviot Hills as a good prospect, and, when Robinson and his family arrived in Nelson in 1856, Robinson set about applying pressure on both the Nelson Provincial Government and John Caverhill to be allowed to take over Caverhill's lease. There was no legal basis by which he could do this, but the Nelson Commissioner of Lands agreed, swayed no doubt by the £10,000 deposit that Robinson slapped down. It was for this kind of flamboyant big-noting that Robinson earned the nickname 'Ready Money Robinson' amongst his fellow colonists.

Caverhill agreed to relinquish his lease on Cheviot Hills, provided Robinson buy him another property over on the Waiau. This became known as Hawkswood. Robinson then proceeded to build up Cheviot Hills into one of the largest and most valuable sheep runs in the country. John Caverhill still had Motunau, but around the same time he negotiated to sell 10,000 acres, including the Black Hills block, to another newcomer, a Manxman by the name of George Henry Moore, who had started aggressively buying up land in the area. Moore had already bought a decent chunk of Teviotdale, some of Charles Sidey's land, the better part of Birch Hollow (Doctor's Hills), and had bought the freehold of the land that Robert Waitt and his business partners were farming as squatters — around 58,000 acres, all told. It was no longer worth Waitt's while trying to farm what was left, so he relinquished the rest of the lease to Moore. By the end of the 1850s, George Henry Moore could put even 'Ready Money Robinson' in the shade. He had built up a holding of nearly 150,000 acres, of which over half was freehold, and at its peak Glenmark Station ran 90,000 head of sheep.

GEORGE HENRY MOORE was born in Billown on the Isle of Man on 12 October 1812. From boyhood, he was close mates with his cousin, Robert Quayle Kermode. Robert's father, William, who had been a merchant and a seaman all his life, first travelled out to Van Diemen's Land (as it was then known) aboard a ship that he owned (named the *Robert Quayle*) in 1819; he returned a number of times, and had already acquired substantial land interests and a founding directorship of the Bank of Van Diemen's Land by the time he emigrated in 1826.

About two years later, his family, including his son Robert and Robert's mate George Moore, joined him and helped him turn his 6,000-acre property, Mona Vale on the Salt Pan Plains north of Hobart, into a showcase of colonial agriculture.

Top Glenmark
Station, Waipara. **Bottom left** George
Henry Moore. **Bottom right**
William 'Ready
Money' Robinson
of Cheviot Hills.

Everyone must have got along pretty well, because in 1839 George Moore married Anne Kermode, William's daughter, Robert's sister and George's own distant cousin. Although four children were born, the marriage wasn't a success: two of the children died very young, and Mrs Moore seemed to prefer the social whirl of Sydney to the bucolic splendours of Tasmania. The couple agreed to separate, and the surviving children — Anne and William — were packed off back to relatives on the Isle of Man.

In the early 1850s, Australians looking for pastures new were reading good things in the newspapers about the land to be had in the newly opened up regions of New Zealand. There was a minor land rush, with the new class of well-heeled Australians who arrived waving cash about and buying up big swathes of land becoming known as the 'shagroons' by those who lacked their means and could only look on enviously.

George Moore was one of this tide, dispatched by the Kermodes to have a look-see at Canterbury land. He arrived at Lyttelton in 1853, and, as he described it, anxious to dodge the groups of men who were hanging about trying to hawk land to speculators and dreamers walking down the gangplank, he set out before dawn the day after his arrival 'with a woollen plaid around his shoulders, a stout stick and a little food in his pockets', walked over the bridle path from Lyttelton and struck out northwards across the Canterbury Plains. He was ferried across the Waimakariri River around noon by a Maori woman in a canoe, and managed to get across the Ashley, Saltwater Creek and (after a meal at Leithfield) the Kowai. As evening was falling, he crossed the Waipara, and spent the first night in what would soon become the heart of his pastoral empire, sleeping in metre-high tussock. He passed most of the next day walking the land, and upon his return to Lyttelton (also by foot), he lodged an application to buy the kernel of what was to become the massive Glenmark run.

The real story was almost certainly a little less picturesque. On at least one of his exploratory forays into the Hurunui, Moore was accompanied by John Caverhill, whom he had met in a Lyttelton boardinghouse. It's doubtful Caverhill was pleased with Moore's 'grid-ironing' Motuanau, but, like the other leasehold farmers affected, there was nothing he could do about it. By March 1854, Moore had purchased 40,000 acres for £17,000 on behalf of a partnership comprising himself, William and Robert Kermode, and a clergyman by the name of Dr John Lillie. The others provided the capital: Moore was to be manager, and the terms of his employment were strictly set out in a contract. One of the homesteads standing on one of the blocks was named Glenmark, reputedly by its original owner, Mark Pringle Stoddard. This name was retained and applied to the whole station.

During the autumn of 1855, the partners sent three consignments of Merinos by the *William Hyde* from Hobart to Lyttelton, and these were driven onto the Hurunui to stock Glenmark Station. And with some of the choice tracts of their land bought out from under them, it wasn't long before Moore's neighbours were offering him the rest. Soon Moore was sitting pretty on 60,000 acres of land, and by 1864 his flock had increased to 64,000 head.

It was New Zealand's age of empires, and small farmers — 'cockatoos', as they were known, doubtless because they flocked together at the stock sales — could only look on in envy as the likes of Moore and 'Ready Money Robinson' sat pretty on their huge estates.

4
AGE OF EMPIRES

I'M IN THE KITCHEN, with an Inia Te Wiata album playing when Nugget barks. I flew Nugget down from my parents' farm in Warkworth just before my last dog Lucy died, on New Year's Day. Here's Russell coming up the hill, five minutes early, as he is for everything, a man after my dad's own heart in that regard. Sometimes, though, that just doesn't work for me, nowadays. I'm often running late, and really try not to.

Russell has been on the place for 64 years — I'm 63, so he's been working Black Hills for longer than I've been on this planet. Today, he's coming to take the back wheel off my ATV for me. I usually check the petrol and kick the tyres, but lo and behold I found the wheel was flat yesterday. Try as I might, I couldn't shift the lug-nuts, so Russell volunteered to do it for me.

Jo Stafford and Gordon McRae are singing away in the next room, now. It will be ABBA, or Amici, or something like that next. I have to have noise in my house. There's a difference between loneliness and aloneness, and music — music and friends, good friends who stick with you — are a big part of the difference. I was thrilled when Russell turned 80 and his children asked if they could have a function here at the restored Black Hills cottage. I was absolutely chuffed, and we had a wonderful celebration. Local poet Christine Benson wrote a poem for him, 80 verses for 80 years, just as she wrote a 100-verse poem for Nan's hundreth birthday.

My dog Nugget and I go out into the yard to meet Russell. We swap a few words and he ambles off to have a look at the ATV quad bike. Everyone should have a Russell.

IT'S HARD TO GET MUCH OF A SENSE of what George Moore was like. He had a bit of a reputation for misbehaving with his young female staff, though. In 1868, a former worker, Angus McKay, brought charges against Moore for 'debauching' his daughter, Eliza, when she was under the age of 21. The matter was settled out of court.

Described as tall, 'commanding-looking, straight as a gun barrel', with a long white beard and piercing blue eyes, Moore was certainly a hard-headed businessman, and probably a bit callous as well, to both man and beast. He acquired the nickname 'Scabby Moore' because of the regularity with which his sheep were found to be afflicted by a tiny, pearly-white mite named *Psoroptes ovis*.

Scab, as this disease was known, is a nasty affliction, which causes sheep great discomfort and, in advanced cases, wool loss, wasting and skin lesions (which is where the name comes from). Because of the Australian experience with scab, the New Zealand government was anxious to keep its incidence

Riverina Marsom
Pulsford Heales, Beverley's
grandmother, probably
taken in the early 1890s.

Top Bill and Alice Miles's 50th wedding anniversary, June 1954. Back row: Bill Miles, Ray Viall, Alice Miles, Louie Viall, Olive Phillips, Noelene Price being held by Ona Price, Jean Francis holding Malcolm Francis. Front row: John Viall, Gary Viall, Beverley Price, Ken Francis.

Bottom Beverley aged four (left) and Noelene aged two (right).

Top Beverley when she won her first champion ribbon at a show.

Bottom Mansel on Dinkie Diamond and Beverley on My Pal, at Warkworth Showgrounds, ready for the Waitemata Hunt, 1968.

Above Mansel and Ona Price — Beverley's parents.

Right Frank and Ellen Forrester — Jim's parents.

Beverley and Jim's wedding
at St Andrew's School Chapel,
Christchurch, 14 June 1986.

Top Jim dipping sheep at 5.30am in 1992 — if you look carefully you can see sheep's eyes shining in the dark on the right.

Bottom Later the same day.

Top Blade shearers clipping Corriedales the old-fashioned (and silent) way. Seven shearers plus shed hand.

Bottom Jim with champion Hereford cow and calf, Christchurch Show 1996.

Top The restored cottage. Jim's grandmother was born here in 1892.

Bottom Beverley at the Hawarden Ewe Fair in front of the pen of BlackHills Heatherdale Corriedales.

to a minimum on this side of the Tasman. But given the flood of sheep coming in from Australia, and the widespread movement of stock around the colony itself, it was always going to be an uphill battle. The Deans boys were struggling with it as early as 1848, and the *Lyttelton Times* mentioned an outbreak in 1850.

The government passed a Scab Ordinance in 1849, which allowed for the appointment of sheep inspectors in each province, the restriction of movement of stock, the branding of infected animals with the letter 'S', and the imposition of stiff penalties on anyone who failed to clean up an infection once it was detected. The most effective measure to control scab was quarantine, and the swift slaughter of infected animals. When a single scabby sheep was found amongst his flock of 6,000, Edgar Jones of Upper Waiau was persuaded by his neighbours to slaughter his entire flock, for which they paid him a modest amount for each animal. Better that than have the whole district declared infected, with all the trouble that meant for moving stock.

There was no proven treatment for scab, but dipping sheep was thought to be effective. The traditional dip used was a solution of tobacco, sometimes with nasties like arsenic, sulphur and saltpetre (potassium nitrate) added. On Glenmark, bales of tobacco (grown on the station) were cut up in a chaff-cutter operated by two hands, who spent their whole time sneezing and probably finished the job with a rampant snuff habit! Sheep were made to swim through the plunge dip, and were ducked with a long, forked rod applied behind their neck, before being allowed to scramble out the other side.

It seems as though Moore was pretty lackadaisical in treating his afflicted stock. He was charged with failing to brand 2,000 infected animals in 1857. In 1863, his business partner, Dr Lillie, appeared in court to defend the Glenmark partners against a charge of having 18,000 scabby sheep on the property, and in 1864 Glenmark was fined a total of £2,400 for offences against the Sheep Ordinances. Appearing in court and copping a fine came to be an annual ritual for Moore: he will have got to know the Canterbury Provincial (Northern Region) Sheep Inspectors, William Congreve and Tom Kinnersley Adams, pretty well. But he doesn't seem to have been overly concerned by scab itself, and there were dark rumours that the notoriously scabby state of his flock was deliberately maintained, as it encouraged his neighbours to keep their boundaries secure and discouraged others from trying to buy the freehold of his leasehold land out from under him.

Moore also hit the headlines in 1860 for his cruel treatment of a swagman, one of the itinerant labourers who were a feature of the New Zealand rural landscape until Michael Joseph Savage and the first Labour government created the so-called welfare state in the wake of the Great Depression of the 1930s. One cold, rainy evening in March 1860, an elderly woodworker named Stephen (also known as Henry) Davis arrived at Glenmark and found the boss in the yard. He asked whether he could have a meal and a bed for the night, but Moore refused, and told him to go to the public accommodation house at Weka Pass, 5 kilometres away over the hills or 10 by track.

The man went in the indicated direction, but tried to talk one of Moore's workers into letting him stay the night in his cottage. The worker, John Henry, also refused, having been in trouble with Moore in the past for putting up swaggers without permission. Since it was pitch black and still raining heavily, he advised Davis to shelter in a nearby woolshed. Davis left, and was found dead of a self-inflicted gunshot wound the following day, just a few hundred metres from John Henry's cottage. When Moore was told, he shrugged and sent for the Kaiapoi constable. He made no effort to have the body retrieved or even sheltered, and when the police came two days later, he refused to put his carpenters to work making a coffin.

The jury at the inquest roundly criticised Moore for his callousness, and the Christchurch newspapers were outraged. A public meeting was held to condemn Moore, and he was even burned in effigy in Hagley Park! His defence — that he turned the man away because he smelled of alcohol and seemed the worse for wear, and because Moore already had five other swagmen in residence — was all but ignored. 'Scabby Moore' was described by an editorial in the *Lyttelton Times* as a 'mean, hard-hearted, barbarous and blasphemous man' (the last because he claimed his carpenters couldn't be asked to build a coffin on the Sabbath), and accused him of acquiring so large a property out of a desire both to make money and to remove himself from the society of decent human beings. 'Inside his boundary, humanity has no rights,' the editor thundered. 'He has bought them up with the freehold at so much an acre.'

There was doubtless a bit of envy in this reaction to Moore's conduct: he was one of the richest men in a colony that was populated, in the main, by people who had hoped they had left behind them the obscene inequalities of British and Irish society.

But while Moore was clearly a grumpy, short-tempered and hard-headed sort of man, he wasn't entirely without his good points. He was instrumental in the discovery and preservation of one of the most important collections of moa bones in a swamp on Glenmark in 1857: his men were digging a drain, and, when they alerted Moore to their find, he sent for the great naturalist, Julius von Haast. 'A large American four-horse dray' was filled with 20 boxes of bones in an excellent state of preservation. Some of the skeletons were articulated and put on display locally, but Haast bartered others for artefacts from collections and museums around the world, and these formed the basis of the collection of the new Canterbury Museum.

Moore was a terrifically hard worker in Glenmark's early

days, and L G D Acland reported him to be a man of his word in business: once, when he had engaged a bullocky to do a carting job, he was told that far less feed had been used than had been allowed for in the contract, and the bullock driver had been selling the surplus at a profit. 'Quite right,' Moore answered. 'That were the contract, so he had every right to do it.'

He was also a generous benefactor to many charities and benevolent funds in the Hurunui and wider Canterbury districts. When Glenmark found itself on the route of 'the human tide' flowing between the east coast up and over the Taramakau Saddle to and from the West Coast goldfields in the 1880s, Moore and his staff were noted for their generosity. Indeed, had it not been for his willingness to kill sheep and bake bread, according to one report, dozens of miners returning empty-handed from the diggings would have starved. And in the 1870s, as New Zealand entered a deep depression and the numbers of men down on their luck increased dramatically — visitors to the Hurunui district counted 72 swagmen in a single day in 1879 — Moore and Glenmark earned a reputation as a reliable provider of hospitality. Perhaps George Henry Moore had learned a lesson in humanity a dozen years before.

Those who worked for him, even while they acknowledged him to be a difficult man, were warm in their praise of him. 'He was a very good boss,' one employee recalled, 'as long as your back wasn't straight!' 'He was mean. Everyone knew it,' Andrew Todd, another former employee, recalled. 'And then suddenly a different Mr Moore would emerge. He'd hear that the coachman's wife, say, was expecting another arrival and he couldn't be quick enough in doing something really good for the young family.'

Moore must have had to pinch himself at times. Whatever dreams or ambitions he'd harboured as a boy and youngster back on the Isle of Man, they must surely have been exceeded by the position he found himself occupying by the early 1880s. His backers and business partners had all died by then: William Kermode had passed on way back in 1852, and John Lillie followed in 1866. Moore's mate and sole surviving partner, Robert Quayle Kermode, died in 1870. That left the ownership of Glenmark in limbo: it was technically still owned by the partnership, and the only way Moore could unlock it was to offer it for sale as the firm's surviving partner and purchase it in his own right. So in the summer of 1873, an auction — one of the biggest the nation had ever seen — was held for the various blocks comprising Glenmark:

> This was in March, 1873, at Miles and Co.'s wool stores, the occasion being to decide the partnership account between Messrs Moore and Kermode. The sale attracted the largest gathering of stock and station owners ever seen before, and some from Australia. Messrs Matson and Co. were the auctioneers, Mr H. Matson having allowed time for 'the fortification of the inner man,' as stated by the 'Illustrated Press,' proceeded to business. The first lot submitted was Glenmark, consisting of 35,781 acres of freehold and 11,500 acres of leasehold, with 25,400 sheep. After a bid from Mr Moore of £65,000, there was some lively bidding between him and the Hon. W. Robinson, of Cheviot, till amid much applause, it was knocked down to Mr Moore at £85,000. The next

lot, Deans Peaks, consisting of 4,099 acres of freehold and 7,500 acres of leasehold, with 5,000 merino sheep, started at £7,000 from Mr Moore, and after a sharp contest reached £13,500, the purchaser being Mr Frank Courage. A block of 3,959 acres at Waipara was bought by Mr Moore for £6,500. The Black Hills, 34,670 acres, held under depasturing license, and 12,500 sheep was next put under offer, and was acquired by Mr Moore for £13,500. The Doctors Hills, 32,306 acres, under lease and 84 acres of freehold, with 12,500 sheep, after fast and furious bidding was also secured by Mr Moore for £14,750. Fifty acres at Weka Pass he secured at 30 s per acre, and one hundred acres at Saltwater creek at £10 10s per acre . . . Messrs Kermode and Moore's properties at that sale realised £186,574.

In 1881, Moore commissioned Samuel Farr to design and supervise the construction of suitably grand buildings for the station. The bald, extravagantly bearded Farr had arrived in Akaroa after a nightmarish voyage out from England in 1849 and, with 40-odd others, chose to stay there rather than push his luck and proceed as planned to Auckland. It was a happy choice: he went on to become one of Canterbury's most notable citizens and architects, even though he had no formal training. He was the first man married in the region, when he wed Mary Ann Pavitt on 15 June 1850, using a ring that Farr himself made from a half-sovereign.

Besides designing some of the lovely stone buildings that graced Christchurch city until the 2011 earthquake — Cranmer Court was one of his, as was St Paul's church; Merivale Manor on Papanui Road is still standing — Farr was one of those who was instrumental in spreading Englishness throughout Canterbury. He was a vigorous member of the Acclimatisation Society, and helped stock the rivers and streams with trout and salmon, as well as bringing bumblebees to New Zealand. And he was a staunch and proselytising Anglican, who founded the first Sunday schools in the settlement.

On Glenmark, Farr designed a monumental stables for Moore, an imposing, gable-roofed L-shaped structure that enclosed a courtyard on two sides and incorporated a granary, a smithy, coach houses and a long loft that contained fodder for the 50 horses housed there (with a name plate for each horse over its stall). It also featured such extravagances as a dovecote and a bell-tower, but while the stables still stands (and has a Category I Historic Places Trust listing, because it was one of the first substantial buildings in New Zealand to be built in concrete), these fancy features have been removed.

Nearby, a small, wooden building with gothic windows and Farr's trademark gables went up — the Station Lodge — and another larger, gabled building with dormers and prettily crenellated eaves was constructed for Moore's manager, Tom Johnstone at that time.

Of course, Moore reserved the grandest flourish of all for himself. Farr designed for him 'a palatial house, with garden and artificial water worthy of a highly refined ambition', as a newspaperman admiringly put it, easily a rival for the 40-room pile that 'Ready Money Robinson' had erected on Cheviot Hills. Perched on a scarp overlooking the outbuildings and realised in brick and concrete, the house was begun in 1881 and completed in 1888, at a total cost of £75,000, a vast sum for the day. It was a sprawling kind of affair, with gables and arched windows and a courtyard surrounded by a wall, and a two-storeyed storehouse with faux battlements that were at once a nod to the great houses Moore will have gaped at as a child in Billown on the Isle of Man, and an assertion of his status as a New World laird. Inside, the furnishings were elegant in the extreme, as a contemporary source recalled:

> The floors throughout were covered
> by heavy pile carpets, and the fittings
> were of oak, even in the maids'
> bedrooms. So were the doors of
> oak: the windows plate glass: the
> sideboards and chiffoniers from a
> Paris Exhibition: the blinds imported
> from Florence (and woe betide the
> maid who allowed the sun an entry):
> and the urns that graced the garden,
> made by Doulton.

When it was finished, Moore is said to have remarked: 'This house is so beautiful, I don't care if I ever cross the Waipara again!'

Partly, all this luxury was bound to have been due to Moore's desire to impress the lady in his life — his daughter, Annie, whom George had brought back out from the Isle of Man in 1867 to keep house for him and stand as heir to his fortune, after an attempt to so cultivate his son had failed. Annie was then 19 and, as a source remembered, a lively young thing who was the focus of much attention from the eligible young men of the district. George put the kybosh on all such advances, assuring his daughter that her suitors were only interested in her money. To compensate, Annie surrounded herself with pets — dogs, 154 pet lambs, at one time, all wearing coloured

Interiors of the original
Glenmark homestead.

ribbons, exotic caged birds, and so many ducks, swans and peacocks that it took 25 dozen loaves a week to feed them.

Moore lavished capital and attention upon his property, too. 'Great energy was shown in regard to cultivation, tree planting, water supply, and subdivision . . . the ideas of a gentleman who had lived a rough and strenuous life,' a newspaper reported, declaring Glenmark to be the best run in the colony. Not only that, but Moore's property was by now the largest: he was running nearly 100,000 head of sheep and he employed well over a hundred people. From the windows of his new house, he was king of all he surveyed.

He must have been pinching himself, alright.

WORKING GLENMARK would have been hard yakker. The size of the stables wasn't just conspicuous consumption (although there was a fair element of that): horses would have been essential to every aspect of the business. Musterers and boundary riders would have used hacks; draught horses would have hauled all manner of agricultural implements, including ploughs for crops of grain, turnips and rape on the flats; fine-limbed carriage horses would have been yoked to a range of vehicles suited to the occasion, because, even on Moore's rare forays off the property, horse-drawn carriages, carts and traps were the principal modes of transportation. Moore kept at least one strong and level-headed horse down by the Waipara River ford: in 1879, the *Lyttelton Times* noted admiringly that the Glenmark mails had got across the flooded Waipara 'through the efforts of Mr Moore's famous river horse, Jerry'. The arrival of the railway at Weka Pass at the beginning of the 1880s changed things somewhat: Moore had property down near Ashburton, and when the pickings were slim at Glenmark he would rail stock down to the greener pastures in mid-Canterbury.

Most hands on Glenmark were engaged on a casual basis at 20 shillings per week, with a bonus of 5 shillings per week payable during the busy times, and especially good men were kept on. Of course, shearing and dipping were the busiest times, as the whole, vast property had to be mustered. Unlike most properties, Glenmark had a large woolshed where shearing was done. Unlike modern woolsheds, where everything is noise and bustle — electric shears, the radio, the shouts of shearers — even with a full gang of 100 hard at work, the Glenmark shed was quiet, as the shearing was all done with blades.

Moore was reckoned by his contemporaries to be an old-fashioned stockman. He wouldn't have anything to do with the new-fangled drafting gates that were beginning to be used. He believed these knocked the stock about too much, and accordingly he had no time for any shepherd who was too lazy to lean over and lift a sheep out of the race, 'and thought nothing of drafting 300 lambs a day'. Andrew Todd remembered the personal interest he took in the work of his shearers:

> He inspected every shearer's work
> as he passed. Detecting a cut, he'd
> stamp with his stick and nearly scare
> the shearer out of the port. Fifty
> such inspections he made — for he
> stopped at every man — and then
> walked out of the shed right beside
> me. My! I'll never forget the remarks
> that filled the air as he left.

There was adversity to face, too. The battle with the old pests was largely won by 1880: the marauding packs of feral dogs that had been the scourge of the early runholders had been all but eradicated, wild pigs were kept under control with a bounty paid for each tail, and scab was finally on the way out, even on Glenmark.

You have to remind yourself that there were no fences on most sheep stations in those days, except perhaps for a ring fence around the odd paddock. Before the invention of light, high-tensile steel wire, iron wire was available but impractical. It wasn't galvanised, merely coated with black paint; it was fully a quarter-inch — 6.35 millimetres — thick and couldn't be strained. Even fencing staples lay in the future for much of Glenmark's history. Instead of stapling wires to posts and battens, they used to bore the posts and thread the wire through.

That made boundary riders among the most important men in New Zealand farming — solitary sorts of fellows who didn't mind spending weeks riding over the hills without seeing another human being. Any sheep straying too close to the boundary was given a sound fright with a stockwhip, and after a while they more or less got the message. Inevitably, though, flocks got mixed up, and at every muster, the waifs and strays were drafted out and penned for collection by their owners. Of course, the whole business of scab made the mixing of stock a big no-no.

In the early days and before the railway was pushed through, all hands on Glenmark had to be gathered to muster stock away from a corridor through which others drove their flocks in transit. This was an enormous job given the size both of the property and of the Glenmark flock. The Canterbury region was largely free of the mite by the 1880s, but it wasn't until 1894 that the national flock was declared scab-free. But no sooner was victory over dogs and scab reported, than battle was joined with other pests, most notably rabbits.

Of course, with two major rivers and several temperamental streams crossing the property, floods were common. Julius von Haast reported being flooded out while working on the moa bone pits on Glenmark. In 1874, a Christchurch court was told that George Henry Moore would not be attending one of his annual hearings for keeping scabby sheep, as the Waipara was in full flood (and presumably Jerry the river horse was unable or unavailable to help, unlike in 1879, when the Waipara flooded again).

Big snowfalls were frequent events, and in 1867 there was a savage dump that caused major stock losses. Yet then, as now, anyone wishing for drier weather needed to be careful, lest they got what they wished for: drought is common here in the rainshadow of the Alps. And when things got too dry, the countryside was chronically prone to going up in flames. It didn't help that the railway line that had been pushed through from Weka Pass north in 1883 was plied by steam trains that belched showers of cinders into the grass and gorse; small grass fires were a common occurrence, but on 8 March 1886 one of these got well out of hand. Over the next couple of days, a wall of fire swept over 50,000 acres of Glenmark Station, and by the time it had burned itself out, it had killed an estimated 10,000 sheep. A reporter who travelled up by train found himself passing for nearly half an hour through a post-apocalyptic landscape, with bare, blackened earth studded with the still-smoking corpses of dead stock. The fire was doubtless a setback for Moore, but it was nothing short of a disaster for another man, a farmer on a far more modest scale by the name of Perrott.

BORN IN SOMERSET, FRANK PERROTT arrived in New Zealand in 1864 with his wife, Harriet, aboard the *Tiptree*. Frank travelled out as part of the New Zealand government's assisted immigration scheme, and his trade was listed as 'bootmaker'. He settled in Rangiora, where he acquired a paddock or two and ran a few head of sheep and cattle, and a horse or two as well. Life doesn't seem to have been easy, and by all accounts Frank was a cantankerous sort of character. He was brutal with his own family. One of his sons remembered that Frank would be bent over his bootmaker last in his workshop while his small children laboured in the vege garden, but even while he was cutting leather and hammering soles he would always have an eye cocked in their direction. If anyone so much as stood up to stretch their aching back, he would amble out, hands behind his back as though merely strolling into the garden. But when he came within range, he would produce a cane and lay about the boy who had dared slack off.

In 1872, he found himself facing a magistrate on charges of having struck a bailiff who was seizing some of his horses under a distress warrant. Soon after that, he began a running battle with his neighbour, his son-in-law who also seems to have been a cousin, Henry Perrott. Frank sued Henry for a sum of money in damages in 1878 and lost, but Henry unwisely counter-sued and Frank won. A couple of months later, Frank was up before the magistrate again, this time on charges of using abusive language towards Henry, who had allowed his sheep to stray onto Frank's land. Called as a witness, Henry's exasperated son told the court that both men were as bad as one another.

In 1882, Frank sold up and moved to the Waipara district, where he rented a farm variously known as Cracker Creek or Karaka Creek on 1,400 acres that fronted the rather grandly named Main North Road. This block was said to be an outstation on Glenmark, and to have been named after 'the native shrubs that grew there' by its first owner, T S Mannering, although there is an alternative story that says it was named after the dry flax flowers that the drovers used to collect there for use as improvised stockwhips, which they called 'crackers'. I've never managed to grow a karaka there. Mannering bought the lease from George Montagu (the Sixth Duke of Manchester, who was a member of the New Zealand Company but who never travelled out to take up his land), freeholded it in 1872, and sold to William Parkerson a few years later.

William and Edward Parkerson were the sons of one of Christchurch's first surgeons, Dr Richard Parkerson, who owned land at Sumner. William Parkerson had purchased Mesopotamia Station from Samuel Butler in 1863 (and on-sold it the following year); settling at Karaka, he had become a stalwart of the Waikari community — a member of the Roads Board, the Licensing Board, the Anglican Church, the Education Board and, of course, the cricket XI. He went on to buy the Lakes Station up near the Hurunui headwaters. His brother Edward was a silent partner in the Karaka deal. He lived in Sumner, where he was accountant to a law firm.

It was probably the Parkersons who had the modest homestead built, and the stables and granary at Karaka Creek in the mid-1860s, from limestone quarried on the property and from cob (made from moulded mud and straw). The new tenants, the Perrotts, moved into the little house and named it Stoneycroft, and Frank plunged himself into the business of improving the land around it.

Life for the Perrotts seems to have been no less a roller-coaster than it had been back in Rangiora, and Frank doesn't seem to have learned to make it easy for himself: in 1884, he unsuccessfully sued the Kaiapoi Woollens Company over some unspecified grievance. There will have been celebrations when his son, Frank Henry Perrott, was married in September 1885, but they were shortlived. In February 1886, Frank Perrott senior was offering a handsome (£160) reward for anyone who returned to him a roll of banknotes that he had lost somewhere on the Main North Road between Karaka and Glenmark on 23 February. Perhaps he had been carrying that money to settle a debt on behalf of his son, for a couple of weeks later, on 4 March, Frank Henry Perrott was declared bankrupt. And then four days later still, it must have seemed as though the world was ending, as fire raged over Karaka Creek, destroying most of Frank Perrott's fencing and all of the grass seed that he had been about to use in further improving his pasture.

For a little while before this disaster, Frank had begun complaining of chest pains and shortness of breath. Then one day in the middle of the winter of 1888, he set off just after 10 in the morning to walk his boundary; his body was found slumped against the fence about two miles south of Stoneycroft by a pair of men named Williams and Nelson. The verdict returned was:

> ... the said Frank Perrott, by
> visitation of God, in the natural
> way, of the disease and distemper
> aforesaid [ie, heart disease], and not
> by any violent means whatsoever to
> the knowledge of the jurors, did die.

Frank Perrott was 66; he wasn't the first and certainly would not be the last man to die of a heart worn out with the strain of trying to farm the Hurunui. I still have the coroner's report today.

Harriet Perrott hung on in there for a while. Poor thing, she managed to fall out of a dray only a month after her husband had died, and broke her arm. But Frank had managed to sow a good 300 acres of the farm in English grasses, and Harriet advertised 150 of these for rent for grazing in 1889. She must have had hopes of going it alone. The Parkersons may have had other ideas: the whole property was advertised for rent the following year:

> RENTAL of 1400 acres (Karaka Creek); about 300 acres is grassed; improvements include stone house and outbuildings, ring fence and subdivisions. Apply Benn & Co., Amberley.

The next time the property was mentioned, in 1892, the 1,458 acres of 'Karaka' were being offered for sale freehold by its owners, William and Edward Parkerson. By now, Edward was in deep financial strife: at a meeting of his creditors in August 1893, his share of the loss on 'a property at Waikari' was stated to be £348.

Karaka wasn't the problem, though. Edward was bankrupted, and the following year found himself in hot water over allegations that he had embezzled money from his employers. He was charged with misappropriating £1,050 from Harper and Co., although the true amount was later determined to be close to £18,000. Never mind he had evidence that one of the principals in the firm, Leonard Harper, had authorised his creative accounting: he was convicted and sentenced to two years' hard labour.

WILLIAMS AND NELSON, the two men who found Frank Perrott's body in his paddock that day back in 1888, were swagmen. Like Edward Parkerson, and perhaps for much the same reasons, New Zealand was going through very hard times in the late 1880s and early 1890s, a period that historians call 'the Long Depression'. Plenty of people who were living hand to mouth were contemplating the ruins of the dream that had brought them out to New Zealand in the first place — the dream of being 'their own man' on a bit of land somewhere, an impossible notion in the United Kingdom and just as unlikely in New Zealand while men like 'Ready Money Robinson' and 'Scabby Moore' had most of the prime real estate locked up in their large estates. As depression hit, the big men, the shagroons, the squatters with capital to tide them over, survived; the 'cockatoos' — the small farmers, most of whom were 'highly geared' (as we say these days) — were ruined. The levels of resentment were running high.

In 1890, John Ballance became Premier of New Zealand, leading the nation's first organised political party — a loose coalition of like-minded people, the Liberals — to power. Ballance had long been an advocate for getting more people onto the land, but, whereas others thought the way to satisfy this 'land hunger' was to extract more land from Maori by hook or by crook, Ballance felt it was better achieved by settling existing areas more intensively. This had been his agenda when he was Minister of Lands in the Stout government from 1884, and he had sought to achieve it by introducing progressive income and land taxes to try to make it unattractive to the owners of large estates to carry on. In a hard age, Ballance was a soft soul, it seemed, who entertained other modern and dangerous ideas, too, such as women's suffrage!

Even so, the man whom Ballance appointed Minister of Lands was inclined to go even further with regard to 'the land question'. John 'Honest Jock' McKenzie had seen at first hand the evils that were done in Scotland during the so-called 'Highland Clearances', where tenant farmers were driven off the land by aristocratic landlords intent on replacing the kind of mixed subsistence farms that supported many families with large-scale sheep farming that supported very few. Given that many of New Zealand's citizens were fellow refugees of the massive upheavals that the clearances had occasioned,

McKenzie's call for the large estates to be 'burst up' was a popular one.

McKenzie himself was in favour of abolishing freehold over rural land altogether, and instead creating a system of perpetual leases. But he had to work within the constraints of what was politically acceptable. He introduced a Land Bill in 1891 that sought to limit the amount of land an individual could own, to oblige them to improve it within seven years of purchase, to ensure that ownership was strictly 'one man, one run' (all intended to prevent land speculation) and to bar married women from taking up land (to stop men using their wives as cover for their own speculations). He had relented somewhat by the time the measure passed into law: married women were entitled to a half-share. Alongside it, he introduced a Lands for Settlement Bill that would empower the government to buy up private land for subdivision.

The first of New Zealand's sheep empires to suffer this fate was the country's second-largest, Cheviot Hills. William Robinson died in 1889, leaving the station to his five daughters. Struggling to maintain it under the burden of the debts it had racked up, they eventually decided to sell to the government; after nearly a year of negotiation over the price, the property was sold for £304,826, and carved up into 54 farms of small, medium and large extent.

By the end of 1894, 178 families — 900 people — were settled on the former Cheviot Hills Station, compared with the 100-odd who had worked it five years previously. The voting public were impressed, and it seemed as though it was only a matter of time before attention turned to Cheviot Hill's bloated neighbour, Glenmark.

GEORGE HENRY MOORE'S decline and fall is like something Shakespeare would have come up with. Whatever foibles he'd had as a young man, he had acquired some genuine eccentricities in advanced age, not the least of which was the kind of unhealthy attachment to his material wealth that self-made individuals who have known real hardship often exhibit. He had directed Farr the architect to design both the Glenmark manager's house and the mansion itself without back doors, and with heavy front doors and gates that were religiously locked at 10pm every night. A visitor to the mansion noticed that there were mirrors disposed about the entranceway that enabled all comings and goings to be watched. Talk about a siege mentality: Moore had made his pile and he was damned if anyone was going to take it away from him.

In 1889, Canterbury was rattled by an earthquake that toppled the spire on Christchurch Cathedral and also shifted the chimneys on Glenmark mansion. A bricklayer who was called to assess the damage suspected that some of the bricks had moved, and recommended stripping the roof to make a better inspection. Moore refused to allow this. Then, just after noon on 23 January 1890 a young woman, who was on the household staff, was with the parlourmaid in the room they shared getting changed for lunch when they heard 'a noise — a rumbling. This seemed to be above us in the roof, and it grew louder and became alarming.' Moore himself came upstairs and saw her preparing to mount a ladder into the attic with a bucket, ready to use water stored in tanks up there for precisely this emergency — to fight what she presumed was a fire.

'That's a man's work, child,' Moore said. 'Come downstairs with me and call them.'

As they went outside, Moore looked up, and the girl saw the colour drain from his face. The entire front of the upper storey was ablaze.

'My house!' Moore cried. 'My house is gone! My house is gone!'

Farmhands ran to the rescue, but there was nothing anyone could do. Annie Moore ran inside and managed to fetch photographs and jewellery from her upstairs bedroom, and was desperate to save her canaries. Even when the lead sheeting on the roof over the front door was melting and dripping like rain from the eaves, she dashed repeatedly in and out to fetch her birds, despite the urgings of the domestics and the farm staff. She let them perch in a Wellingtonia tree in the forecourt, but it caught fire, too. She was quite badly burned trying to rescue her birds a second time. Meanwhile, Moore simply stood by and wrung his hands and moaned: 'My beautiful house! My beautiful house!'

The house was a total loss, the newspapers reported, and

the valuable furniture, plate, pictures, tapestry, and cabinets of treasures were all destroyed. There was no insurance, and when asked whether he would build again, Mr Moore quietly said 'No.' 'Well,' said the writer, 'will you do as Cheviot owners have done — sell to the Government?' After lunch, in the manager's house, with Mr Wynn-Williams and Mr Withnal, of Miles and Co., Mr Moore produced a copy of the offer which was then made, tendering the Government the whole of Glenmark, minus the homestead block of 4,000 or 5,000 acres, at a very reasonable price — about £4 [per acre].

117

Confident that he was finished at Glenmark, Moore relocated his household to Sumner instead, where Annie started another flower garden, having, according to one source, 'some of the finest specimens of orchids, ferns, etc in New Zealand'. She was a supporter of the SPCA, and once again surrounded herself with dogs, pet lambs and 'birds of all kinds', including swans.

Moore, meanwhile, was losing his eyesight. He had given over management of Glenmark to his staff, and was living the life of a gentleman. He was keen for the government to buy him out, but he wasn't going to let the place go for a song, and, while the two parties were reported to be close to agreement in 1896, negotiations were testy and the rhetoric flowed freely on both sides. Member of the House of Representatives (and future Premier) Richard Seddon told the House that 'if a train takes three hours to pass through one man's estate, it was time to break it up'. Moore wrote to the newspapers in reply, stating that the 'down train' took less than an hour to traverse Glenmark, including a stop at Waikari. The 'up train' took a little over an hour, as it climbed 600 feet on its journey. Readers of the paper probably thought quibbling over how many hours it took for the train to cross Moore's land was beside the point. A petition reported to occupy a piece of paper 50 feet long was presented to the House of Representatives in July 1897, containing the signatures of those who had missed out on the Cheviot Hills land ballot and wanted the government to lay its hands on Glenmark, no matter what it cost.

But negotiations broke down completely after the government solicited a valuation of the land that came in well short of Moore's own estimation. There were 71,787 acres of freehold in question, ranging in quality 'from agricultural to purely pastoral': the government's assessor thought that the land was worth £4 2s per acre improved, and £3 17s 8½d unimproved. Moore reckoned the improved land was more like £4 5s. Well, he would say that, the government assessor retorted. Moore, he said, had a long history of inflating the value of his land.

Shakespeare would have enjoyed the fact that a hailstorm featuring hailstones 'the size of pigeon's eggs' hit the Omihi Valley soon afterwards.

Meanwhile, since the disastrous fire, Moore had been quietly breaking up his estate all by himself, selling off small blocks of land here and there at prices ranging from £4 4s (for the Deans Block) to a whopping £12 per acre: the government assessor grumbled that these prices were jacked up between vendor and purchaser to influence the government.

By 1900, Moore was completely blind, and it seems his long-suffering daughter saw her chance: in September, Annie secretly married Dr Joseph Henry Townend, who was a widower. George never learned of the marriage, and nor did he notice when just two years later Townend died and poor Annie was bereaved. George Henry Moore — 'Scabby Moore' — died at Sumner on 7 July 1905, aged 92.

His estate, needless to say, was massive, but the authorities soon noticed that it didn't seem to reflect the proceeds of all the selling down that Moore had been doing over the preceding decade. They traced the missing money to Annie's bank account, where Moore had been systematically stashing it. Annie was duly summoned to the Supreme Court to answer charges of being party to an attempt to evade death duties. The Supreme Court found that Moore's campaign of land sales and the gifting of the proceeds to Annie was indeed an attempt to evade death duties, but it also found that there was nothing wrong with it. Annie escaped penalty. With close to £1 million sitting in her account, she was the richest woman in New Zealand.

One of her first acts was to commission the construction of a church in her father's honour on Glenmark: St Paul's church was consecrated in 1907, and the first baptism held there on 27 July of that year.

Shortly after her father died, Annie Quayle Townend bought a property in Fendalton Road — part of the old Deans run — from Fred Waymouth, the manager of Canterbury Frozen Meats. It was named Karewa, but Annie renamed it Mona Vale after her grandfather's estate in Tasmania; she also inherited her father's large, twin-gabled house in Sumner. She clearly inherited many of her father's characteristics, too: when she found herself in a dispute with the Sumner Borough Council, she had the house lifted and shifted bodily by traction engine up the track that was to become Dyer's Pass Road to a new site out of Sumner's jurisdiction.

Annie was a staunch supporter of the Canterbury A&P Association, and remained a dedicated gardener. She donated a begonia house to the city's Botanical Gardens — its replacement still bears her name — as well as collections of a number of flowering plants. She was a generous but low-key benefactor to a number of charities, and when she died in May 1914 she handsomely recognised her many friends and staff in her will, along with relatives by blood and by marriage and even a distant cousin back in Billown, Isle of Man.

Mona Vale and its beautifully landscaped and planted gardens were eventually purchased as a park by the City of Christchurch. The gorgeous homestead operated as a café and visitor centre until the 2011 earthquake knocked it about. It's still closed for repair.

BY THE TIME ANNIE TOWNEND DIED, Glenmark was a shadow of its former glory, now covering just 11,500 acres. The government purchased it lock, stock and barrel in December 1914, and announced that it would be carved up into blocks and sold by ballot. Interest was enormous: no fewer than 11,000 people lodged a total of 15,000 applications for the 25 sections, which ranged in size from 11 acres to 897 acres.

All of this was going on over the boundary fences of Karaka and Black Hills, both of which were now in the hands of a 66-year-old Scotsman who had land interests all over the Canterbury region. Robert Adam Forrester was one of the new kind of South Island sheep farmers, the 'cockatoos', the small men. But he was destined to be a small man in quite a big way.

5
THE SMALL MEN

I'M CHUFFED!

We've had a busy time of it at the Canterbury Show, as we always do. We were showing 14 animals this year, and the logistics of getting them down there is always a bit of a trial. Jude was a big help. We base ourselves in a caravan for the week, so that had to be shifted on-site as well. And I'm always flat out during the show itself. I'm on the Show Committee, so I'm always called on for steward's duties. Most of my fellow committee members are men, and they do the same range of stuff. Most people have no idea of all that goes on behind the scenes. I'm also on the Cattle Committee, which often conflicts with my various other activities.

This year, besides all of my steward's duties, I was showing my lovely big coloured Merino ram. He was up against 109 others in his section and, would you believe it, he won! I was pleased with that result. But because he'd won his section, he then went into the champion of champions competition, and blow me down he came third, behind Jim Sidey's white Corriedale and a white Merino from Stevensons' over at Cheviot. Not a bad effort for the black sheep in the business! Who knows what the various generations of Forresters who preceded me on my land would have thought. But he's an impressive boy, with a real presence about him, a real look-at-me mana.

It reminds me of the time — 1995, I reckon — when the Canterbury A&P Show had a sheep class where the complete unit (the ewe, her lambs, and their wool) was converted into a dollar value. I sent a side sample to the organisers, as required, and they accepted my sheep, despite it being black. I told husband Jim, and he was dead against my taking it to the show. But I snuck her onto the trailer. Jim was pretty dark about it, and once we were at the showgrounds, he wouldn't let me do the feeding, in case anyone worked out that he was affiliated with this funny-coloured animal. A mate of mine, Sharon Paterson, did the feeding. Well, I beat all the white sheep! I was embarrassed, so I left it to Sharon to find Jim and tell him.

'Jim Forrester! You won't believe what your wife has gone and done,' she said.

'Oh crikey,' groaned Jim. 'What's Beverley done now?'

'She's only gone and blimmin won with her black sheep!'

We received a special prize, and boy was Jim chuffed!

The following year, the terms of entry had been re-worded, keeping coloured sheep out.

It's always a relief when the show is over, much as I enjoy the buzz around the place and the camaraderie. There's just such a lot of work getting organised for it, and, because I put my rams out a month earlier these days than I used to do, things are more complicated. The rams used to go out around 25 March, which meant lambing began mid-August. But for one reason or another — getting caught up in Fashion Week was one, and the fact that my neighbour's rams were getting to my ewes before my rams had a chance — I shifted things back. It's turning out not to be ideal, because you leave the sheep you're going to take to the show in wool, and you shear them straight afterwards. If the lambs are a month older when they have that first shear, that's a month's worth of wool you lose. Life is all about these balancing acts.

SOON AFTER SHE ANCHORED IN Lyttelton Harbour on 24 November 1862, many of the passengers of the 1129-ton ship *Chrysolite* published a letter in the *Lyttelton Times* to her master, Captain Duncan McIntyre, thanking him for his care and competence in delivering them safely and in relative comfort from one side of the world to the other. They likely meant it: unusually for an immigrant voyage, not one casualty was recorded, even if it was a little slow — at 106 days from Gravesend to Lyttelton, it took over a month longer than the *Chrysolite*'s first voyage in the colonial service the previous year, which lasted only 74 days.

The vessel had the wind against her for much of her run down through the Atlantic to the Cape of Good Hope, and, after a spell of favourable weather in the Southern Ocean, a 'complete hurricane' overtook her a few hundred miles from New Zealand and carried away some of her sails. Even for passengers who were well used to the motion of the ship by then, and who had even begun to believe she would get them to their destination alright, it must have been a terrifying experience. And then, as if to mock them, she was becalmed for four days on the east coast of the South Island.

But the only serious mishap occurred when she was at

journey's end and dropping anchor. When her second anchor was let go, a passenger by the name of Richard Sadler had his ankle fouled in the hawser, and his lower leg was torn clean off about three inches below the knee. It was sometime before he could be loaded onto the harbour-master's launch and taken ashore for whatever passed for medical treatment back in the day, accompanied by his wife, Catherine, whom he had married only a short time before their long voyage began. The leg was amputated above the knee.

Sadler was from Middlesex and was just 21, and his occupation was listed as 'labourer'. Like most of the rest of the *Chrysolite*'s complement, he was travelling with the assistance of the Canterbury Provincial Government to start a new life in the colonies. The *Colonist* newspaper reported approvingly that 'he bore very acute sufferings with manly fortitude', and urged its readers to show their usual generosity towards the unfortunate man. It seems they did: crew and fellow passengers of the *Chrysolite*, along with various others, had raised nearly £34 by Christmas 1862, roughly three times what it cost him for his outward fare. Still, his situation was pretty desperate: government assistance of any description ended the moment an immigrant tottered down the gangplank. Once ashore, it was sink or swim.

BETWEEN 1853 AND 1870, New Zealand's non-Maori population rose from 30,000 to 250,000, with Canterbury receiving most of this influx. And of those who came to Canterbury, by far the majority were assisted immigrants and the lion's share of these were Scots.

The *Chrysolite*'s human cargo was therefore pretty typical of those of the migrant ships at the time. Of the 295 souls aboard, most were Scottish, many English, quite a few Irish, a handful were Welsh and there was a Manxman, four Germans and a couple of Belgians. As you'd expect, given that they had been handpicked by the Canterbury Province's representative, most were engaged in 'useful' trades, with agricultural skills to the fore: 45 of the men were 'farm labourers', 14 shepherds, and 6 ploughmen. There were 50 single women aboard, a product of the colonial government's enthusiasm for the idea of evening up the gender ratio of the young nation, which had attracted far more men than women in its frontier days. Many of the married couples aboard had young children travelling with them, the youngest being month-old Maria Chambers. There were 7 infants, and 32 children under the age of 12, all told.

Robert Forrester paid £13 towards his fare. William, a couple of years older than Robert, was travelling with his wife, Margaret, their five-year-old daughter, Jean, and one of the *Chrysolite*'s babies, three-month-old William. He paid £32 10s. Robert and William's older brother John was travelling with his wife, Agnes, and two daughters, six-year-old Isabella and two-year-old Catherine; he paid £39 to bring his family out. John, William and Robert, all the sons of a more senior William Forrester and Helen Adam of Kilsyth, Scotland, weren't the only Forresters aboard: there was a James Forrester, whose father was a different William, and whose mother was Elizabeth Gilchrist. Just to confuse everything, the two Williams senior were likely cousins.

Who knows, these days, what it was like taking ship in the circumstances endured by the Forresters and their families and their fellow passengers. The *Chrysolite* was fairly new, having been built in New Brunswick in 1858. Her first-class accommodation was pretty impressive, by all accounts, with roomy cabins, a stateroom, a saloon and even a smoking room 'where devotees of the nicotian weed may indulge in their favourite luxury', as the *Wellington Independent* enthused when the vessel visited Port Nicholson after discharging her Canterbury-bound passengers.

Not that the assisted immigrants, travelling in third-class or steerage, got to indulge many luxuries: they were accommodated in the 'tween decks which, while roomy, were essentially a cargo hold, with hammocks slung for beds and blankets hung to offer a modicum of privacy. The pleasant smell of lanolin from her homeward-bound cargoes of wool would soon have got lost beneath the odours of 250-odd unwashed, seasick, homesick and (in many cases) just plain sick bodies living in close proximity — just bearable when conditions allowed promenading on deck, but hellish, you'd have to think, when inclement weather kept everyone below decks.

And never mind the smell. There was the movement of the ship, unsettling at best, sickening for most of her landlubber passengers, and occasionally straight-out terrifying. And the noise — the rush of the sea, the creaking, cracking, groaning of timbers, the rattle of sea boots and the muffled cries and shouts of the crew, all unfamiliar and hardly likely to help those new to ships and their habits to relax. But that's how the majority of the ancestors of New Zealand's Pakeha population got here. It's a measure of how strong the lure of fresh fields and pastures new was to young people in depressed and over-populated nineteenth-century Britain.

THE FORRESTERS were an old Stirlingshire bunch, a sept of the MacDonald of the Isles branch of the Clan Donald. The name first appears in Stirlingshire in the twelfth century: it

Robert Forrester from Kilsyth.

derives from the office of the keeper of the King of Scotland's forests. Over the next few hundred years, Forresters were very much in favour with the crowns of both Scotland and England, fighting with distinction in most of the notable battles in this period of history.

In the seventeenth century, Sir George Forrester, who had married into another noble family, the Livingstons of Kilsyth — he wasn't the first Forrester to marry into this family, and strategic matrimony is likely what brought Forresters to Kilsyth in the first place — was made Baronet of Nova Scotia by England's ill-fated King Charles I, and made a peer of Scotland, his title being Lord Forrester of Corstorphine. As he had no male heir, George's title passed to his son-in-law, James Baillie, who as second Lord Forrester made the mistake of declaring his loyalty to Charles when Oliver Cromwell was in Edinburgh, which saw his estates (apparently quite substantial) sacked and pillaged. Poor old James took to drowning his sorrows in local alehouses, apparently much to the disappointment of his ambitious mistress Christian Hamilton (who also happened to be his niece). In 1679, she called him out of the pub, exchanged angry words with him and murdered him with his own sword. While she managed to escape from prison dressed as a man, she was re-apprehended a day later and beheaded on 12 November 1679.

There were other Lords Forrester after James, but the family's fortunes were now firmly in decline. The family seat, the Castle of Corstorphine, fell into disrepair and was sold in 1701. It has long since been demolished, although the Forrester association with the site is commemorated in several street names in the Edinburgh suburb of Corstorphine, and in the Corstorphine Parish Kirk, which was erected by the Forresters and dedicated to St John the Baptist in the fourteenth century, and which still stands. The Forrester title itself was merged with another in the English peerage in the early nineteenth century.

Little of the family's fame, glory or wealth had devolved on John Forrester's branch of the family, which in the early eighteenth century was living in the little village of Kilsyth, around 40 kilometres northeast of Glasgow on the main road to Edinburgh, part of Stirlingshire in those days, but now in North Lanarkshire. The village was once known as Monybruch, but the name Kilsyth was firmly established by modern times. There were lots of Forresters about the district, many of them working as farmers or (believe it or not) vintners. It is believed that John Forrester married Isobel Gillespie in the middle of the eighteenth century, and that they had five children, three sons and two daughters. One of their sons was William Forrester, whose own son William married Helen A Hamilton Adam — curiously enough a descendant of the murderous Christian Hamilton: this is a rare instance, it seems, in the history of Scottish clans — with their notoriously long and vindictive memories — where bygones were allowed to be bygones!

Kilsyth had become a coalmining town since the advent of the Industrial Revolution, but William Forrester was recorded in the 1841 Census as a weaver by trade, and as a cotton-weaver on the marriage certificate of one of his children. He and Helen had five children, and it was three of these — John, William and Robert Adam — who migrated to New Zealand in 1862. The other son stayed behind, and may not have survived into adulthood. Katharine is thought to have emigrated to the United States sometime after 1869 (when she was still in Scotland). Living with the Forresters when the 1841 Census was collected were the McNair brothers, John and Hugh. After her husband, William, died in 1843, Helen Forrester married Hugh McNair and had two further children. The fact that she named one of these James suggests that her earlier son James had passed on.

ROBERT ADAM FORRESTER was born in Kilsyth on 9 January 1842 and baptised the same day, as was the custom when it was far from uncommon for children to die in infancy. However, Robert escaped that fate and seems to have thrived, and was a little under two months shy of his twenty-first birthday when the *Chrysolite* delivered him and his two brothers to Lyttelton on 24 November 1862.

Britain in general and Scotland in particular were in the grips of a severe economic depression in the middle of the nineteenth century: the Scottish agricultural revolution known as the Highland Clearances had displaced much of the population of the Highlands, and the masses of the displaced had placed enormous pressure on the lowland regions as well. To young men growing up, the land of their birth must have looked finished, and devoid of opportunity. For the Forrester boys, like thousands of their countrymen, the chance to start a new life with the assistance of the New Zealand colonial government would have been a godsend. On the other side of the ledger, Scottish shepherds were much sought-after in New Zealand: their skill with sheep and tough country was ideally suited to the New Zealand hill and high country. Much of New Zealand's prosperity is down to the influx of Scotsmen from the middle of the nineteenth century.

Robert's brother William died a little over a year after their arrival, aged 25; while William entered New Zealand as a 'farm labourer', his occupation at the time of his marriage was coalminer, and it could be that the health effects of this dangerous business had caught up with him. He was buried in the Addington Cemetery at Sydenham. Margaret, his wife, was pregnant at the time; after the baby was born — a son she named James — she took her three children back to Scotland in March 1865. She later married another native of Kilsyth, William Shaw.

Upon arrival in Lyttelton, Robert's brother John (also known as Jack) went into partnership with a man named Miller. Together, they secured a 'scavenging' contract for the Lyttelton Municipal Council — collecting rubbish and running a night-soil cart. The pair also landed an asphalting contract or two, before successfully applying in 1865 to lease a block of 52 acres of council land designated as botanical reserve. Their stated intention was to work this land as a market garden, although they seem to have gone in for dairying instead, as John soon began supplying milk to the town of Lyttelton. In one of those circularities of history, this land was part of the Purau estate once farmed by the Greenwoods, who went on to farm Black Hills as part of Motunau Station.

John and his wife, Agnes, had seven children besides Isabella and Catherine, whom they brought out from Scotland. John died in 1902 at the respectable age of 70.

James, the cousin of William, John and Robert, followed in his father's footsteps and set up as a butcher in High Street in Lyttelton, first in partnership with a man named Tom Gunnell, and then with another by the name of Dixon. In 1865, he married Mary Stirling, and also successfully applied for a publican's licence. Soon afterwards, he and Mary travelled back to Kilsyth, where their two children, Jeanie and William, were born; when they returned to New Zealand in 1870, they settled in Colombo Street, Sydenham, where James ran his own butchery and, as a little entrepreneurial sideline, also had an agency for the Union Insurance Company. The butchery flourished, and eventually ran to two shops (one with a slaughterhouse and commercial freezing plant attached) and employed 10 men and 12 horses. James played an active part in local body politics and was three times mayor of Sydenham Borough. He died in 1923 at the age of 81.

ACCORDING TO FAMILY LORE, Robert Adam Forrester arrived with all his wordly possessions in a carpetbag — I still have it, and it's on display in the restored cottage — and 2s 6d to his name. He first took work as a barman when he landed in Lyttelton. But soon afterwards, he was hired as a shepherd on William McRae's rugged Glens of Tekoa station in the upper Hurunui Valley, literally in the shadow of Hurunui Peak. The landscape here will have appealed to him as a fair facsimile of home, and certainly the life seemed to suit him and his set of skills.

After two years on the Glens of Tekoa, he moved on to Montrose Station down the road at Culverden. Montrose was approximately 42,000 acres at this time, and belonged to Sir David Monro, who was Speaker of the House of Representatives for 10 parliamentary sessions, and the grandfather, incidentally, of the man who is credited with having introduced rugby football to New Zealand. Robert was Monro's head shepherd on Montrose, and he served in that capacity until around 1877. In 1876, Monro sold Montrose to William Rutherford, one of the seven Australian-born brothers who between them owned and farmed most of the Amuri district in the late nineteenth century. William Rutherford persuaded Robert to stay on for a year, but by now Robert was just about in a position to start making his own way in life.

Farm business would regularly have taken Robert out to Saltwater Creek, a settlement about 12 kilometres inland from Kaiapoi. A man named James Ashworth had based himself here in the early days, and served as the principal carrier of goods inwards and outwards from the stations of the Hurunui district. He used to carry limestone from Karaka and Black Hills as a backload. There were hotels, general stores, a smithy and a post office there as well. Coaches, wagons and drays used it as a staging post, and small coasters even made their way up the creek to the landing stage for a time. In fact, Saltwater Creek even had its own steam navigation company, subscribed to by most of the big station owners of the Hurunui, including George Henry Moore; the company got as far as buying its own steamer, intended to take out the district's wool clip. Forseeing bright things for the port (named Northport), one entrepreneurial soul even bought up tracts of the riverbank to sell as sections in what he expected to be a bustling township. But the provincial government was lukewarm about maintaining a port in the shallow Saltwater Creek estuary, and when the great flood of 1868 washed most of the settlement down the creek and out to sea, the writing was on the wall for Northport. When the railway reached Sefton, about five kilometres distant, much of Saltwater Creek's commerce relocated there.

For some reason, though, Robert Forrester's diary continued to record visits to Saltwater Creek. Blow the mail and the stores: it appears Robert had his eye on a transaction of a different kind altogether. On 22 May 1876, Robert married Louisa Burrell of Saltwater Creek at The Manse in Christchurch, the Reverend Charles Mason officiating. In a bit of a departure from the terse, scribbled notes about the weather, the movements of stock and the state of assorted boundaries, the entry for 3 June in Robert's diary reads: 'Came home to Montrose with long haired mate.'

LOUISA BURRELL was just a year old when her family travelled out from England as assisted immigrants aboard the *Isabella Hercus*. They arrived at Lyttelton on 4 January 1856, after a harrowing voyage — not at all atypical of immigrant voyages — and were immediately placed in quarantine. There had been a fever aboard. There had been deaths.

Louisa's parents, 26-year-old labourer George Burrell of Reigate, Surrey, and his 22-year-old wife, Sarah, were accompanied by George's older brother Edward, his wife, Emily, and their one-year-old daughter, also named Emily. Edward was 29, and was also described as a labourer. Whatever hopes he had of bigger and better things for his family on the far side of the world didn't even survive the voyage. Poor little Emily was one of the four infants who died en route; the same number of adults didn't make it. And Edward's wife may have been one of the 10 passengers who were hospitalised on arrival: she died only three and a half months later. Somehow, Edward managed to pick himself up and carry on, and he was soon working as a brickmaker in Rangiora.

George, meanwhile, settled in Ferry Road, Papanui, and worked as a bushman for a while. He first went to the Ashley district when he successfully tendered for a contract to form a road through the swamp lying between Saltwater Creek and what was to become Sefton. While he was working the contract, George built a little sod hut on land leased from Mount Grey Downs Station, which belonged to the redoubtable Sarah O'Connell, who had been doing famously, farming it on her own account since her husband, Major Edward O'Connell, had left her a widow in 1853. By all accounts, she had probably run the show before Edward died, as one contemporary source described him as 'a bit of a goose'. Sarah O'Connell, by contrast, was warmly written about by everyone who knew her. One woman who enjoyed her hospitality described her as 'not very ladylike, but good-looking, with something very pleasant about her manner, and, as her name may suggest *very* Irish'.

As George Burrell's contract neared its end, he bought land at Sefton. It's likely he went into partnership with his new son-in-law, Robert Forrester, to buy another block about 10 kilometres away at Fernside, a little to the west of Rangiora. When gold was discovered in the Taramakau and Buller rivers in the mid-1860s, George left wife Sarah in charge of the farm at Sefton and tried his luck 'over the hill'. He was then drawn to the goldfields further south around Greymouth and Hokitika when they opened up. After his return to the Ashley district, he and Sarah shifted to the Fernside property in 1877. Robert and Louisa joined them a year later, with a babe in arms. William Adam Forrester had been born on 30 March 1878.

Soon after William's birth, Robert and Louisa had left Montrose, going to stay for a while with Jack Forrester, whose established position at Lyttelton probably offered the most comfortable prospects for the new mother and baby. Robert had lately acquired some land, in partnership with Thomas Greig of Christchurch: a block of 796 acres at South Cape in the neighbourhood of Rotherham in the Amuri district, part of the St Leonards Station, recently subdivided.

And in a note in his diary in late 1878, Robert makes his first reference to a block in the Hurunui that he is expecting to take over that has yet to be surveyed. This was a section of the Black Hills, which George Henry Moore had carved out of Glenmark and was subdividing, selling the land on a deferred-payment basis. The exact nature of the deal Robert had struck to acquire this land is unclear, but his diary entry indicates that another party owed it to him, perhaps in lieu of cash payment for work done. He seems at this stage to have been biding his time until he could start his new life as a farmer on his own land, picking up casual work: for the Elder brothers in Christchurch, at the Glens of Tekoa again for a couple of months, on his father-in-law's property at Fernside, digging drains and clearing the swampy, flax- and bracken-covered ground.

In February 1879, a month after the birth of his second son, Robert Adam, Robert sold his interest in the Fernside property, along with the cattle it was carrying. He, Louisa and babies Robert and William shifted to Rangiora for a month, and then headed north to the South Cape property to farm it. Photographs show an attractive (if small) timber homestead and outbuildings on a pleasant flat in the Waiau River Valley, with established exotic trees already mingling with the straggling remnants of lowland podocarp forest. This is where Robert and Louisa lived. Louisa's brother William did a five-month stint assisting them in 1880, on a wage of £1 a week. Another of Robert's brothers-in-law, George, helped with the ploughing.

At the same time they shifted to South Cape, Robert noted in his diary that he had received 'half his land at Hurunui'; the rest was due to him within two years at 9 per cent interest per annum. Robert received the balance of the land in March 1881, presumably along with a handsome cash settlement in respect of the interest. Two days after he recorded the transaction in his diary, he noted that he had bought Tom Greig out of the South Cape farm. He named this property Kilsyth. Nine months later — the timing may or may not be significant — Robert and Louisa's third son, George Albert, was born. There would be two more boys, James Kilsyth in 1883 and Matthew Burrell in 1885, then a daughter, Ellen Jane, in 1887.

Robert was one of six men who bought up blocks in the Black Hills subdivision. His was a small section: when he paid rates to the new Waipara Roads Board in 1885, it was on the basis of a freeholding of just 280 acres. His occupation was still described as 'labourer'. But in 1892, Robert bought the Karaka and Perrotts blocks adjacent to the Black Hills off the Parkerson brothers, and the family moved there around the same time, living first in a little wooden house (now gone) on Karaka, and then shifting to Stoneycroft on Perrotts. Yet another child — another girl, Sarah Agnes — was born on 10 May 1892. Robert seems to have systematically bought out his neighbours on the Black Hills themselves. By 1901, his holding there had grown to 1,350 acres.

Robert and Louisa Forrester
with their children at Kilsyth,
Rotherham, c. 1890.

BY THE TIME their eighth and last child, John Archibald, was born in December 1894, Louisa would have been all but worn out. She was 39 years old; and while her three eldest boys were all but grown up, and Nell, her elder daughter, at seven was likely to have been useful to her, the entire burden of Victorian-era home-making otherwise fell on her shoulders.

Between them, Louisa and Nell would have done the cooking and washing for all 10 members of the family — both hot, heavily manual tasks with an open fire for cooking and a copper for the laundry. Then there was the business of tending to baby John and 17-month-old Sarah, as well as cleaning the house and making and mending clothes for everyone. Even if they could afford it and were minded to buy it, bread was delivered to Waipara only once a week in those days; Louisa doubtless made her own, and not only milked the house cows, but also churned the butter. In season, she would have laboured, red-faced with sleeves rolled up, over a preserving pan. And every now and then, when stock needed shifting urgently, say, or some other operational aspect of the farm needed attending to, she may have worked alongside Robert out on the hills.

Trips to the outside world, even to Waipara, would have been few and far between, and required the hitching up of horses to the trap or taking a saddle horse. On each occasion, Louisa would have sat with pencil and paper and laid out her plans for the next few months as carefully as any ship's quartermaster preparing for a voyage. It was her responsibility to lay in enough stores to see her crew through the journey of a season.

It's little wonder that when Louisa was photographed at Karaka in 1896, flanked by Sarah and Nell and with little John on her knee, her eyes slid away from all the photographer's fussing about with his black hood and his little tray of magnesium powder. A slight, wry twist to her mouth and her downcast gaze give her a look of exhausted resignation.

$$\Rightarrow\!\!\longrightarrow$$

LOUISA'S FATHER, GEORGE BURRELL, died in 1898 at Waikawa in Southland, where he and Sarah moved in 1890 after letting the Fernside property. It would have taken a few days for this melancholy news to reach Louisa, and it would have been out of the question for her to have attended the funeral, due to the length of time it would have taken to travel by steamer, even if her family could have spared her the time away. The newspapers described George as 'the beau ideal colonist.' Then, a little under five weeks later, Sarah Burrell died, too. The papers were again fulsome in their praise of a woman whom they described as 'an ever ready helper to her plucky pioneering husband'. She was just 67.

Meanwhile, though, the Forrester family was prospering. Robert continued to add to his land holdings, as opportunity arose. He must have acquired some sort of interest in Jack's property at Lyttelton, as he received regular rent payments from his brother. In 1899, he bought 1,816 acres at Waitohi Peaks, formerly part of the Horsley Downs estate up near the headwaters of the Hurunui. He acquired paddocks here and there — a five-acre spread at Waikari, and two-acre paddocks at Hurunui and Rotherham. In 1903, he added the 101 acres formerly farmed by John Olson at Waikari, and picked up Block 8 in the subdivision of a swathe of Glenmark adjacent to Karaka. He even attempted to buy the Manapouri Station in Southland, which he scouted on a visit in 1904. In 1905, he bought a 4,874-acre property he called Narrowdales at Milburn in Otago, and the Upper Waiau run formerly belonging to Edgar Jones on the northern boundary of the Glens of Tekoa. His last purchase was a plot of land in Sawyers Arms Road in Papanui: perhaps he was intending to retire there.

Of course, Robert's firm expectation was that his sons would populate and operate his pastoral empire. He shifted them around from property to property. His older three boys worked closely with him on Karaka and Black Hills, but Bob was given much of the responsibility for running Narrowdales after the acquisition of that property, and George was largely responsible for running Kilsyth. However, Robert's oldest son, Will, went to work as a teamster for a while at Longbeach Station near Ashburton. Matt was enlisted to drive his younger sisters and his brother to Hurunui School, as they were too young to handle the pony and trap; once his duties in this respect were discharged, Robert set Matt to work on the Jones property, to which Matt commuted on horseback, a long, rough ride up the course of the Hurunui.

Mustering a flock of Corriedales,
November 1964,

6
NICELY FITTING GENES

PEOPLE IN THE CITY probably think the world's problems are solved and big deals made around shiny tables in boardrooms, but in rural New Zealand just about anything of real importance gets discussed and decided on leaning over the rail of a saleyard. It was by way of a muttered aside as they leaned on the rails at the Hawarden Ewe Fair that Doc Sidey learned from Jim that he was to be an executor of Jim's will, along with Ross Johnson. Neither of them thought much more about it at the time, of course.

Doc and I are there right now, leaning on the rails of the Hawarden Saleyards, looking the sheep over. I'm not really in the market, but I like to keep my eye in on these things. I still go and help sometimes when the trucks offload the bulls up at Molesworth Station in January. We used to send a few bulls up, because, unlike most Herefords, Black Hills' stock were reared in hill country and adapted well to the rugged country on Molesworth. I love the dust and excitement, and the camaraderie between the truck drivers and us farmers. Most of them have been doing this work for years, and that makes for a bond that's hard to define. It's a business relationship, but it's one that has deepened into tradition, into a way of life. I think that's what I love about it so much.

The Black Hills Hereford stud is still on the books, and I'm still a registered Hereford breeder, but I'm not active in that field anymore. I just don't have time. My stock — I have a few head — are up on Doc's place, now. The only cattle I have about the place are my lovely Highland cattle. My parents had a few Highlanders along with their Angus, and I got my first pair in 2006, lovely shaggy red beasts with those wide, spreading horns, which are a big hit with the bus tours. People love feeding them. I called them Hairy and Harry. My next pair were called Woolly and Wally. I was offered a couple of in-calf cows in 2012, but they were black. It must be Dad's prejudices rubbing off on me — he used to wince when he drove past a paddock where cattle of various breeds rubbed shoulders — and I felt the same. I liked my red boys, and I didn't want the blacks in there spoiling the look.

Doc is an old mate. He was at boarding school with Jim and Ross Johnson and that lot, and he's married to Jim's cousin, Janice. In many ways, Ross and Doc remind me of Jim. It's that whole 'still waters run deep' thing. They're quiet men, but with a lovely sense of humour, and both the ability and the inclination to study things from all angles before they make a pronouncement.

Jim's dad, Frank, was a great one for the sales. Shortly after I mated up with Jim, Frank was presented with a beautiful silver salver by the stock agents at the Hawarden Ewe Fair to mark his fiftieth consecutive year of attending the fair with his stock. Not only was he a regular contributor, but he was a big contributor, too, one of the biggest. For Jim, the annual driving to market was a massive logistical exercise, and I was lucky enough to be involved for a few years.

Beginning before dawn, we would drive the mobs — about 1,200 ewes in total, with one person and their dogs at the head of each mob, and another one or two at the rear, whistling, barking, and progressing on — along the road, sometimes encountering another mob at an intersection. One of you would have to hold your lot until the others were safely out of the way, and then you'd get going again. It was too far to go from Black Hills to the saleyards in a day, so we'd camp in our hogget paddock along Bentleys Road there, and then it'd be up by 5 the next morning to get going to make the saleyards by 8.30. We were usually the first there, the mobs were drafted up, and then we were sipping coffee and having breakfast with the stock agents as the sun rose over the Black Hills to the east.

It was a huge community day, and a real captive market for people selling all kinds of stuff. The orchardists used to do pretty well out of it, too. Most of the wives in the district would see their husband roll in at the end of sale day, sunburned and probably a little merry from the after-match, with a big box of fruit he'd bought. 'Here you go, love,' he'd beam, and she would sigh and muster a wry smile, kissing goodbye to whatever plans she'd had for the morrow and resigning herself to a day spent slaving over a preserving pan.

The Hawarden Ewe Fair is still a big deal, although they split it into two these days, with the younger animals one day and the older ones another. People come from further afield, but you'll still get a good 22,000 ewes going through. We used to get lots of people coming and staying with us before and after the big day. It was one of the main events on the calendar.

DOC SIDEY'S ROOTS are firmly planted in the Hurunui, too. He and Jan farm Glenovis, a bit to the west of Mason's Flat and about five minutes' drive from Hawarden. They moved there after Doc's mum, Cath Sidey, shifted away to Mallochvale. Cath was born in Christchurch in 1911, trained as a nurse and worked for Mrs Ginny Sidey. And the job had what you might call a fringe benefit. At some point, she caught the eye of one of Ginny and David Sidey's two sons, Harry, universally known as Mac. They were cousins by marriage: Ginny was Cath's brother's mother-in-law. But love will find a way.

Besides Doc, Cath had three other children, another two boys and one girl. Mac was a keen dog-trialler and deeply involved in the Canterbury A&P Association, serving as president in 1958. Hosting the trials and the A&P meetings required the laying-on of a fair bit of hospitality, and, of course, that burden fell on Cath's shoulders. All the same, she was active in the Horsley Down Anglican Church, the Red Cross, the Women's Division and the Historical Society, as well as in Plunket and the Mothers' Union. After Mac died in 1961, she lived on at Glenovis for a few years. Late in life, she moved to Mallochvale Cottage to make way for Doc and Jan, but returned to live in the little cottage on Glenovis for the remainder of her days. She died in 1990, aged 79.

Ginny Sidey, Mac Sidey's mum, was born at Corriedale, just outside Oamaru, one of the seven children of James and Mary Little. The names of both James Little and Corriedale are pretty illustrious in New Zealand agriculture. In fact, there are places right around the world where you will run across the Corriedale name, and you'd be surprised where in the world you can be where even mentioning James Little can draw a little flicker of recognition.

NEW ZEALAND has always been wonderful country for pastoral farming. But the overwhelming disadvantage we've always had in getting our produce to the world market is our distance from anywhere. These days, technology has found a way around the tyranny of distance, but for the first four decades of the history of New Zealand farming, it was pretty much the sole consideration when it came to deciding what stock you ran on your property. Modern-day dairy farming wasn't dreamed of, because, apart from the nearest town, there was no one to buy milk, butter or cheese from you. You couldn't reach more distant markets, because your product would have spoiled by the time it got there.

It was the same with meat. People raised cattle in a small way to meet local demand for beef, and there was an export trade in hides. And while there were large numbers of sheep, they were pretty much exclusively wool breeds, with the Merino understandably dominant. You could export wool and hides, but, in the absence of any method of preserving meat for the three-month journey back Home, there was only so much mutton New Zealanders could munch their way through, especially once the gold rushes and all those hungry miners had gone. Mutton was a staple, as you readily appreciate when you consider the way in which we even used mutton to replace poultry on festive occasions: a joint of mutton stuffed with bread and onion was known as 'colonial goose'. So believe it or not, anything in the way of sheep meat we produced over and above what the local market could consume was waste product. Sheep carcasses were boiled down to produce tallow, which could be shipped abroad and used in manufacturing, mostly candles. But it wasn't uncommon for farmers to muster up mobs of sheep that were past their prime and simply drive them over a cliff.

Two developments changed the whole face of New Zealand agriculture. One was technological. The other was genetic.

DR GEORGE MURRAY WEBSTER came to New Zealand in the 1850s to take up the position of Deputy Inspector-General of Hospitals, having previously served as a military doctor in Canada and risen to the rank of Surgeon-General in the British army. He apparently intended to buy land in Canterbury, but was unable to do so, for whatever reason. Instead, he settled in North Otago and, in partnership with William Aitken, purchased 20,000 acres near Oamaru from Mark Noble. The land was in two blocks. One, the pair named Balruddery; the other, Webster named Corriedale, after the farm on which his mum was born back in Scotland. The Scottish Corriedale is at Strathblane, a bit to the north of Glasgow, and incidentally less than 20 kilometres from where the Forrester family hailed at Kilsyth. Corriedale was sold in early 1874 to Richard Oliver, who became Minister of Public Works.

Who knows what Webster was thinking when he decided he would stock his property with Romney Marsh sheep? Perhaps he intended them for a Canterbury run, and, by the time he had given up on finding anything in Canterbury and concluded the purchase of the block at Waiareka, it was too late to cancel the order. Otherwise, you'd have to suppose he was badly advised, either about Romneys or about North Otago, or perhaps both. But it was Romneys he wanted, and Webster hired a young Scottish shepherd to procure a mob and bring them out to raise at Corriedale. The young Scottish shepherd was James Little.

James Little was born on Bowbeat Farm, Temple Parish,

Midlothian, on 22 October 1834, the oldest of six children by his father's first marriage. Like so many of his countrymen, the skills that saw him flourish in New Zealand were acquired on the hill country of Scotland. It's not unlike the New Zealand landscape to look at, and it certainly fostered the same class of stockmanship that distinguishes Kiwi farmers.

Little married Mary Jane Telfer in 1863, and two years later he boarded the good ship *Canterbury* with his wife, daughter, and a mob of Romneys — 22 ewes and 2 rams — bound for Lyttelton. They arrived on 18 June, and then trans-shipped by the *Geelong* for Oamaru, arriving there 'in first-rate condition', as the newspapers reported, on 28 June.

Romneys had already shown up in New Zealand. The first confirmed importation was by Alfred Ludlam, who farmed at Waiwhetu in the Hutt Valley in 1853. While they were pretty good candidates for the North Island — they thrive in lush pasture and were widely believed to be resistant to the nasties that afflict sheep on swampy ground, like foot-rot and liver fluke — they weren't that well suited to the conditions that Corriedale offered. It doesn't rain that much down that way, and the pasture was mostly native tussock — ideal for picky browsers like Merinos, but not promising at all for Romneys.

Still, it doesn't seem to have held Little back. Although he hadn't laid eyes on a Romney before he took on Dr Webster's mob, he clearly had a keen eye for a nice animal. On the journey out, he had identified a ram and a ewe that he judged to have particular potential. He christened them 'the Duke and Duchess of Kent', and within a couple of years their progeny was picking up prizes at North Otago stock sales.

Nevertheless, Dr Webster's neighbours assured him he was making a big mistake trying to run Romney Marsh animals on Corriedale: a farmer can always count on their neighbours to tell them what they're doing wrong! The disheartened Dr Webster talked it over with James Little, who proposed having a crack at cross-breeding the Romneys and the Merinos, especially if other farmers could be persuaded to build up cross-bred flocks, too, so that the gene pool could be refreshed occasionally.

Stock breeders back in those days probably didn't have much of a theoretical grasp of genetics. Gregor Mendel, the Austrian friar who first described the laws of heredity, was busy at work with his sweet peas at exactly the same time James Little was heaving his way through the Southern Ocean towards New Zealand, and it was a good 50 years after Mendel first published his findings before anyone took much notice of them. But human beings had been selectively breeding farm animals since time immemorial, and good breeders had a 'feel' for what crosses would produce what traits — even if the whole complex business of genetics occasionally confounded them.

It was well enough known that if you put a Romney Marsh ram over Merino ewes, at least some of the progeny ought to exhibit the desirable characteristics of both: the fine wool of the Merino, the hardiness, the fertility, high growth rate and the refined maternal instincts of the Romney. Little's results were promising, and he was soon doing a smart trade in selling Romney-Merino-cross rams.

The prevailing wisdom was that you couldn't 'fix' the character of cross-breeds: in other words, if you mated two cross-bred animals, the results would be wildly unpredictable, and the desirable traits that the parents possessed would be lost or diluted in the subsequent generation. The answer, Little reckoned, was to cull heavily, eliminating animals that didn't have the characteristics you wanted from your breeding flock.

Webster gave Little every encouragement and more ewes. According to Little, he proceeded like this:

> When we docked the lambs, I kept
> what I thought were the best of the

ram lambs to put to 600 ewes. These had a distinguishing mark and were put to graze all the winter with the other half-breds out of Merino ewes. Before lambing, however, they were to be drafted off so as to lamb by themselves, and at docking time they were earmarked and branded and were grazed with the half-breds out of Merino ewes until shearing time. They were then drafted off by themselves and compared with the half-breds. They were shorn separately; the wool was weighed and compared by the wool classers, with the wool from the half-breds and the inbred hoggets. It was found that their [the inbreds'] wool had a slight advantage in both weight and quality, and the sheep had a most decided advantage in conformation. It was found that they also had fewer of the bad points of the Merino as a mutton sheep.

Little's progress was interrupted when Dr Webster died in 1878. He shifted to the Hurunui, to Hawarden, where, in partnership with his brother Robert, he leased a 3,600-hectare property named Allandale from its owner, Henry Gray. The Little boys stocked Allandale with around 4,000 Merino ewes they sourced from neighbouring properties, and with the late Dr Webster's flock of Lincolns, another English longwool breed. James resumed his experiments with cross-breeding, this time using Lincoln rams over Merino ewes.

Funnily enough, similar experiments were happening elsewhere on both sides of the Tasman. Since 1874, the manager of the extensive land holdings of the New Zealand and Australian Land Company (NZ&ALC), Canadian-born William Soltau Davidson, had been tinkering with Lincoln-Merino crosses at the NZ&ALC's North Otago property, The Levels (which, incidentally, was founded by the Rhodes brothers, neighbours and benefactors of the Greenwoods in their Purau days). When the time came to try to assign a name to the sheep that he and James Little (not to mention various Australian breeders) had independently developed, Davidson proposed 'Southern Cross'. In the end, though, it was James Little who was hailed as the prime mover in the development of what came to be known officially as the Corriedale.

While Davidson missed out on his share of the glory for developing the new breed of sheep, he acquired a lasting claim to fame in New Zealand agricultural history all the same. In February 1882, NZ&ALC dispatched a ship named the *Dunedin* from Port Chalmers to London with nearly 5,000 frozen mutton carcasses aboard. Davidson had been instrumental in getting funding for the experiment, sourcing the refrigeration gear and the vessel, and seeing the two put together: he waved goodbye to the *Dunedin* when she sailed, and he was there to meet her, probably a little anxious (and who could blame him) when she arrived at her destination. The hatch covers were thrown off and the cargo inspected, and over 4,000 of the mutton carcasses were still in edible condition. It was a big success, and showed that the refrigerated export of meat right around the world to the United Kingdom was possible.

It's hard to overstate what a big deal this was for New Zealand. Because there were so many sheep being grown in New Zealand and Australia, world wool prices were in steep decline. But thanks to Davidson and the technology of refrigeration, the game had changed. The population of

Great Britain was growing, and so was its demand for food. Suddenly there was a way of getting all that protein that was just going to waste on New Zealand farms to the dinner-plates of hungry Britons.

Of course, if you just happened to be running a flock of dual-purpose sheep — sheep that produce high-quality wool and an excellent carcass — then you were in a position to benefit both from the wool clip and from the new export market for meat. James Little was doubtless encouraged to apply himself to his quest to produce a fixed inbred half-bred sheep with renewed vigour.

BY THE MID-1890S, James Little's efforts at maintaining a flock of inbred half-bred sheep were made that much easier by the fact that several others were doing the same. Besides Davidson's flock at The Levels, a number of other farmers were raising flocks from Merino-Lincoln crosses, and at least three farms — the Ensor brothers at Mount Grey; Watson Shennan at Conical Hills in Otago; and James Little's brother-in-law, Alex Telfer (universally known as 'Sandy' for the colour of his hair), who farmed at Mount Hilton — were doing much the same thing with Merinos crossed with Border Leicesters (another traditional English longwool breed). In 1903, a meeting of the New Zealand Sheepbreeders' Association voted to include a record of inbred half-bred sheep in an appendix to the annual Flock Book. It accorded James Little the honour of heading the lists of flocks.

Little relinquished his lease on Allandale in 1903, and moved his flock to a property close by that he had bought and named Dalmeny Park. He had six grown-up daughters by now: a seventh, Janet — the child who accompanied James and Mary out from Scotland as an infant — had died in Christchurch on 20 June 1890, aged 25. All but one married farmers from the Hawarden district: the odd one out, Ellen, moved to Christchurch, where her husband was a coach-builder. In 1900, James's second daughter, Margaret, married David Sidey, the son of a Hawke's Bay Presbyterian minister, who did some time as a farmhand at Little's Mount Adern property. Sidey had been ballotted a property at Lyalldale, inland from Timaru, as the Liberal Government broke up some of the huge South Island sheep runs. He and his new wife — everyone knew her as Ginny — moved there and farmed Lyalldale for the next 21 years.

In March 1910, there was a meeting of breeders of inbred half-bred sheep, at which it was agreed to form their own association and to seek recognition of the breed from the Sheepbreeders' Association and to apply to the Canterbury A&P Association to have mutton and wool competitions included in the annual show. The Sheepbreeders' Association agreed. After a bit of bickering between Little's supporters and those of William Davidson at regional meetings up and down the country, it was decided to prefer the name 'Corriedale' to 'Southern Cross' for flocks of inbred half-bred sheep based on Merino and longwool breeds.

In 1916, 20 Corriedale flocks were upgraded from an appendix to the Flock Book to entries in their own right. Each flock was now numbered. Since it was a requirement of full Flock Book recognition that a flock have a proven record of 15 years' continuous inbreeding, and since James Little had dispersed his Allandale flock in 1905 and the Dalmeny Park flock in 1916, the New Zealand and Australian Land Company's two flocks — one at Moeraki Estate and the other at The Levels — took pride of place as Flocks #1 and #2, respectively. Flock #5 belonged to James Little's son Henry (known to all as Harry), who had an interest in a number of properties, but who became famous for his work with Corriedales at Hui Hui. Harry was present at the inaugural meeting of the Corriedale Association, and he was also at the association's meeting in March 1917, which drew up the first type standard for the Corriedale breed. This move was likely taken in response to an approach from a group calling itself the American Corriedale Association, which had written seeking a set of judging criteria for US stock shows. By now, there were also several flocks of Corriedales in Australia.

By 1920, there were 75 registered flocks in the Sheepbreeders' Association Flock Book, and no fewer than 20,000 Corriedale ewes. There was a burgeoning live export market in the breed, with flocks flourishing in the United States, Australia and South America, and even one — the only foreign stud to be registered outside New Zealand — in Kenya. Corriedale wool was given its own class by the Colonial Wool Brokers' Association of London in 1923, and the following year the Corriedale Sheepbreeders' Association voted to break away from the New Zealand Sheepbreeders' Association and to incorporate as the Corriedale Sheep Society, publishing its very own Flock Book. The breed had properly come of age.

BETWEEN THE YEARS OF 1878 AND 1916, when he finally moved to Christchurch to retire, James Little made himself a feature of the Hawarden landscape, as prosperous farmers do. He was a member of the local Roads Board and Rabbit Board, and he was active in the local Agricultural and Pastoral Association. He was a competitive spirit, as a charming little anecdote in Tom Burrows's history of the Corriedale breed attests. Apparently, one day when he was en route to Christchurch in his Model T Ford, Little was overtaken by a flash Talbot belonging to one of the Rutherford family from their vast sheep-farming empire at Amuri. Little told his driver to step on it, but the poor old Tin Lizzy was delivering all she could. Little sat tight-lipped for the rest of the journey, and wasted no time in trading in his Ford for a Talbot when he got to Christchurch.

James Little didn't live to see the foundation of the Corriedale Society, but of course he had the satisfaction of seeing the breed established, and it then becoming one of the most economically important in the country. He died on 31 October 1921 at the age of 87. In his own memoirs, he wrote that 'all things considered, I found life too short to get to the bottom of everything'. Well, we all know that feeling.

ON 29 APRIL 1896, James Little and a travelling companion were farewelled with a bit of a do at the British Hotel in Lyttelton. The two men were off to do a tour of inspection of the British end of the frozen-meat trade, and to see if they couldn't pick up some decent Border Leicesters from British breeders. It was a successful trip, as they reported to a group of well-wishers who welcomed them back with another do at Burke's Hotel in Lyttelton in October. James Little's mate on that trip was Robert Forrester. The two men doubtless knew one another socially, given they lived so close. Even if they hadn't, they would have crossed tracks often enough. Robert stood unsuccessfully for the Waipara Roads Board in 1893, and hardly a year went by when he wasn't grumbling to the board about rates, or the need for fences, or his fervent wish that

cocksfoot be planted along the roadside adjoining his property.

On their trip to the Old Country, Robert visited farms, saleyards, the great meat market at Smithfield, and butcher's shops in England, Scotland and Ireland. He visited distant relatives — nephews and cousins named Shaw, Adam and Forrester — and closer ones: his brother's boy, Andrew, was training to be an engineer at Dumbarton, and Hugh McNair, Robert's own half-brother by his mother's second marriage, lived at Lum Hill in Worcester. He also visited his birthplace. Kilsyth seems to have made a gloomy impression:

> I had a look all round Kilsyth it was
> raining in the evening. I met a lot of
> my old chums. I did not know them
> at the Cross. Kilsyth has not altered
> much since I left it [34 years before]
> . . . I had a talk with a lot of the old
> miners there. Wages are very low now
> . . . I had a good look around Burwood
> and there are a good many nice
> houses getting built in it now. Nice
> cottages and gardens.

He paid his respects at his parents' graves — the sad news of their deaths would have reached him out in New Zealand, weeks after the event — and he also visited the old family home: 'I called at Kernloch and had a look round the old place. No alteration about it. I had a look around the old ground I used to go. Not much altered. Uncle's shop still a flesh shop where my uncle used to live. I could find no one I knew there.'

Whether his association with James Little had anything to do with it or not, Robert Forrester was already experimenting with cross-bred sheep by the late 1880s. In 1889, he sold 195 cross-bred wethers for between 14s 9d and 17 shillings per head, and a mob of 35 ewes for between 12 shillings and 12s 6d per head at the Addington sales — appreciably more than others were fetching. In the 1890 Sheepowner Returns, he reported he was running 1,000 head on his various properties. This had tripled by the time the returns were published in 1901.

Like James Little, Robert was a pillar of his community. He did eventually manage to get himself elected to the Waipara Roads Board, and from 1899 he also served on the Amuri County Council. He was a member of the Hurunui School Committee from 1890 to 1898.

Robert took ill in late 1908. He told his diary on 27 December that he had been sick, and that he had lost 10 pounds in three weeks. On 4 January, he wrote:

> At Karaka. Fine day. I was sick. Jim
> sowing turnips and Barney harrowing
> house paddock and barning and Jim
> went cutting oats on No 8 Flat. Matt
> stripping in forenoon and went to
> McCleans for plough afternoon. Dr
> Little and Mrs A. Alden called.

Then another hand takes over the entry. One of his sons has written:

> Dad passed peacefully to rest at
> 8 o'clock.

The official cause of death was given as 'chronic bronchitis'. It's a measure of how well regarded Robert Forrester was that his funeral was attended by 31 trap-loads of people and about 50 others who got there on dray or horseback, besides a good turnout of his own extended family.

ONE OF THE VEXED ISSUES that arises when a farmer dies is that of succession — and it's no different today than it was 100 years ago, except perhaps that families were larger then. Robert Forrester had acquired quite a portfolio of properties throughout the South Island, and he had eight children. Never mind any sentimental attachments any of the children had formed with the land, the various financial relationships were complicated: in many instances, one of Robert's children would take over the mortgage or title of a property and farm it while Robert retained ownership of stock and plant. Some of the mortgages were assigned to Louisa and to Robert's daughters, Nell and Sarah, although their brothers farmed the blocks involved. You can bet there were some pretty fraught conversations between 1909, when Robert died, and 1911, when Louisa died at Karaka on 5 August, aged 57 (her funeral was just about as big as Robert's had been).

After a trip to the Supreme Court, a family agreement was drawn up to sort it all out. The agreement assigned particular blocks of land to each of the Forrester children. Will Forrester, the oldest, farmed Karaka Downs (as a 900-acre block of the Karaka Creek property was known), and rented the 1,300-acre Black Hills from Robert's estate. The balance of Karaka went to Jim. Bob got a block at Milburn, Otago, which Robert had purchased in 1905. George, who had worked in close partnership with Will on Kilsyth, Karaka and Black Hills, received Kilsyth. Matt got a block at Upper Waiau that Robert had purchased in 1899. Robert's will provided for a property to be bought for the youngest of his children, John, who was barely 15 when his father died and not quite 17 when his mum passed over, too, and this was duly done. Louisa's share of Karaka — some 536 acres — passed to Nell and Sarah.

Nell bought her sister out of her share when Sarah was married in 1915. She built the graceful timber homestead that still stands on Karaka, but she didn't end up living there. She married in 1925 (to a Forrester, a relative from Kilsyth, as it happens) and settled in Christchurch. John had sold the property that had been provided for him when he was called up to go to the First World War, and found himself rootless when he was invalided out of the draft by medical inspectors. He bought Nell out of Karaka in the early 1920s, and when he was married to Jessie Black in 1924, he and his bride shifted into the homestead.

Simple, really!

JAMES KILSYTH FORRESTER — BETTER known as Jim — added to the part of the Karaka block he inherited from Robert, and named the expanded property Heatherdale. In the meantime, he married Bertha Holder, the oldest of the three children of Frank and Jane Holder of Horsley Down. Bertha was a nurse at Kaiapoi Private Hospital when she met Jim. They were married on 30 December 1914, and their son Francis — Frank — was born on 9 May 1916. Frank was to become my father-in-law.

Jim fulfilled the time qualifications to register the Heatherdale Corriedale stud in the 1922 Flock Book, where it was designated Flock #52. The newspapers find him fetching top or close to top dollar with his cross-bred stock in the local sales throughout the 1920s. These were pretty tough times throughout New Zealand, of course: the calamity of the First World War had been followed by the equally catastrophic influenza epidemic. And while there was a sort of boom in

James Kilsyth Forrester and
Bertha (née Holder), with
their son Frank (my father-
in-law).

the middle years of the 1920s, this petered out and eventually tipped into the Great Depression. The New Zealand landscape was haunted throughout the period by swaggers — lost souls, for the most part, displaced by loss or misfortune, and eking out a living by carrying their worldly belongings wrapped up in a blanket or a sugar-bag (their 'swag') from property to property seeking casual employment. Some were highly skilled; some were not. Some were hard and willing workers; some preferred to idle along.

The reception swaggers got varied from property to property, too. Some landowners were happy to employ willing men, and a few were compassionate enough to find work for those less fortunate or to put them up for the night even when they were feeling the pinch themselves. Others were brutally exploitative or callously dismissive. Word soon spread amongst swaggers as to where you might find a warm welcome, a square meal and perhaps a fair pay for a day's work — and which places to avoid. In the Hurunui, Heatherdale was one of the properties that earned a reputation for the generosity of its quietly spoken, likeable boss and his kind and hospitable missus.

Poor Jim died young. He was only 48 when he died on 4 October 1931. Bertha was only 40, and Frank was just 15. Courageously, she carried on at Heatherdale, hiring people to help her run the farm. Her family rallied around: the house at Heatherdale was often full of her nieces and nephews. You can bet the local community played its part in seeing her through, too, just as it did 66 years later when I found myself in much the same position as Bertha.

OVER THE YEARS, THE CORRIEDALE BREED had gone from strength to strength and had risen in importance both in New Zealand and internationally. The first international breed conference was held at Lincoln Agricultural College in 1950, and the keynote address was delivered by the President of the New Zealand Corriedale Society, Jimmy Little, the grandson of the breed's founder. One of the two vice-presidents alongside him was his cousin, Mac Sidey, who had been farming Glenovis to pretty good effect since 1941, and who had been serving on the Corriedale Society council since then, too.

Conferences — they call them 'world congresses' these days — have been held at various locations around the world about every five years since that inaugural one in 1950. The 2007 World Congress was held in Christchurch, and one of the highlights was the unveiling of a statue of a Corriedale sheep made from Oamaru stone in a little tussocky garden adjacent to the Rocking Frog Café at Waikari. The plaque on it reads: 'in memory of James Little and those Corriedale breeders whose vision and skill ensured the spread of the Corriedale from this area to all the major sheep rearing regions of the world'. Of course, these days there are all kinds of fancy technologies that they can use to mark genes so that you can pretty much guarantee what traits will be passed from one generation to the next. So it was nice to pause for a while to remember the 'vision and skill' of the long line of stockmen since James Little who made New Zealand's first breed of sheep what it is and was.

The winner of the Supreme Award at that Congress was John Sidey, Doc's brother who farms Strathblane (named after the part of Scotland where Corriedale is) just up the road at Waikari. Between them, those two maintained the family dynasty in Corriedale shows: the Glenovis and Strathblane names grace most of the larger trophies, and the Christchurch

Show honours board as well. Both Doc and John have served on the Corriedale Society council, and have worked tirelessly to promote the breed that their great-grandfather founded. And between them, they have raised four boys who have inherited their dads' passion for the Corriedale breed, too, ensuring that the Sidey name will remain synonymous with the Corriedale for some time to come.

Top The three Sidey brothers, John, Murray and Doc, at the Christchurch Show in 1982, with their Corriedales.

Bottom Beverley's spotted Corriedale ewe with lambs in 2013.

7
HOME ON THE HILLS

WINTER USED TO BE A QUIET TIME for sheep farmers in the Hurunui. Back in the Glenmark days, there was nothing much to do but cut posts and rails on the fine days and play cards as the rain and snow set in. These days, of course, there's not much let-up at all. All my ewes have to be given selenium and vaccinated for nasties like pulpy kidney, malignant oedema and suchlike, and the two-tooths have to be given boosters two weeks later, allowing two to four weeks before the onset of lambing. You have to work backwards all the time to remember where you are.

And, of course, because Jude and Russell are only part-time, I only have them on certain days. I've only just taken Jude on. She lives locally with her little boy, Lyall; her partner works as a truck driver and is spending most of his time down in Christchurch helping with the rebuild. Jude has been deaf since birth, and, although she has hearing aids and can make out what you say if you speak up, she naturally has some speech impairment. But as you so often find with people who have spent their lives dealing with this kind of adversity, she has a ton of spirit and a real independence of mind. I reckon she'll work out.

Jude has another job or two and, of course, she has her four-year-old, Lyall, so she can't come on full-time. Her work situation with me is complicated. You'd think there'd be nothing better than letting a four-year-old have the run of the place while all that farmwork's going on, but you just can't do it these days. Occupational Health and Safety have strict rules, never mind that farm kids pretty much have free rein in their parents' workplace all the time. And Lord knows what whole generations of farmhands, shearers, fencers and so on would make of the laws that now make it illegal to smoke as soon as you pass the front gate of a farm, which is technically entering a workplace!

There's also a whole new raft of rules and regulations relating to the leaching of nitrates and what-have-you from pasture into waterways, and rightly so. It's about time that side of things was tightened up, and I'm all for making sure farming is done in an environmentally friendly manner. But I was reminded of my situation when Federated Farmers and the Hurunui Water Project people summoned us to a meeting with Environment Canterbury a couple of weeks ago to discuss the issue. I went along, of course, but with only one of me on the farm it's hard to get away just like that. You can see why so many farms work only because there's a team — even if it's only a husband and wife — running it. It's just about impossible to make a go of it alone. You can have your whole week planned out and suddenly something like that meeting is thrown in. It's worse when it's social obligations, too. Just the other day, a good friend died. Don Rowe worked on Karaka for donkey's years, and I wanted to go to the funeral, so everything had to be changed around. Another friend's funeral was held at a time I had already promised my mate Helen Heddell two or three months previously that I'd go and help her with their annual bull sale. If there's two of you, you can split up and keep everyone happy. But with just me, you end up doing everything at a hundred miles an hour and spreading yourself pretty thin.

FRANK KILSYTH FORRESTER took over the Heatherdale Corriedale stud from his dad's estate in 1937. He was married at St Andrew's Church in Culverden, on 9 December 1941, to Ellen Black, the second of four children of Arthur and Sarah Black of Culverden, whom he had known from childhood. Well he would, wouldn't he? They were cousins, Ellen's mum being Sarah Forrester, Frank's dad's sister. It was wartime, and petrol rationing was in full swing, which made the prospects of having any kind of honeymoon pretty bleak. But friends and relatives pledged their petrol coupons so that Frank and Ellen could do a road trip to Glenorchy.

It was just their luck, really: New Zealand had lifted itself out of the Great Depression of the 1930s just in time to experience the hardships and labour shortages of the Second World War. But better times were on the horizon. Our unwavering support for the Mother Country in her hour of need gained us most-favoured-nation status, and New Zealand farmers were guaranteed a market for their produce in reconstructing Britain.

By the beginning of the 1950s, we were entering a period of prosperity, boosted by a spike in the demand for wool caused by the Korean War, in which the United Nations armed forces were fighting in a bitter climate and in desperate need of warm clothing, but also because of the invention of aerial topdressing. Superphosphate — an artificial fertiliser made by treating rock phosphate (usually guano, or fossilised bird droppings) with sulphuric acid — had been around for close to 100 years by the time Frank took on the farm. But applying it was a laborious business: it was done by hand, with workers walking backwards and scattering the powder or granules from a sack worn in front, as though they were feeding chickens.

Trials of using aircraft to spread super were first conducted before the war, and the results were promising. But it wasn't until 1948 that a series of systematic trials was jointly conducted by the Air Force and the Department of Agriculture, largely prompted by the lack of manpower. It was a period of full employment. Jobs of all descriptions were plentiful and labourers in demand, so who wanted to spend their days wandering the hillsides throwing handfuls of acrid super about the place? The results of the trials were again very promising, and on 27 May 1949 a demonstration aerial application was made just down the road from Christchurch at Tai Tapu, on a property belonging to Sir Heaton Rhodes. The applicator was Airwork, which had been using Tiger Moths to drop poisoned carrots on rabbit-afflicted country. The same apparatus — a Tiger Moth with a hopper built into the front seat — proved very effective in delivering super.

New Zealand was pretty flush with skilled ex-Air Force pilots, and with Tiger Moths as well, as these had been the trainer plane of choice for fighter pilots. So there was a general scramble to set up aerial topdressing companies, and, on the ground, farmers started scouting for suitable spots on which to bulldoze an airstrip. Jim went in with his neighbours to place a strip on a ridgeline just west of Mount Alexander. From a distance, it hardly looks long or flat enough for a plane to take off from, let alone land on, but that was the skill of topdressing pilots — and still is. Aerial topdressing transformed much of

New Zealand, and the Hurunui is no exception. Stock-carrying capacities rose dramatically, and doubled in some cases, compared with what you could get using traditional methods.

THE OLDEST OF FRANK AND ELLEN'S FIVE CHILDREN, James Kilsyth, Jim — my Jim — was born on 9 October 1943. Helen Frances came along three years later, Christine Wendy (known as Wendy) three years after that, and Rosemary Margaret three years later still. Another girl, Susan Joy, was born in 1957 but, sadly, died in 1959 of meningitis, aged just two.

Ellen would have kept busy! She was well used to domestic labour, because she had left school after two years boarding at Christchurch Girls' High to help her mum keep the Black family house and gardens. But for all the technological advances that had been made that improved the lives of rural women — most notably the motorcar: Ellen was a good driver — running Heatherdale was still very labour-intensive. Cooking was still done on a coal range, and there were often others to feed besides her big family: stock and station agents, shearers in their season, farm cadets, family friends and relatives. Washing was done in a copper, and both cooking and doing the laundry required coal and wood to be hauled. Like most fastidious farmers' wives, Ellen insisted on scrubbing her dining room table, kitchen bench, doorsteps and verandah with sand soap and polishing the kitchen floor lino daily. Over and above this, for many years she had her mother-in-law living in the same house, after Frank's dad, Jim, died.

But Ellen somehow found time to be active in the community, too: she got involved with the Girl Guides in 1965 and was District Commissioner until 1973, whereupon she became

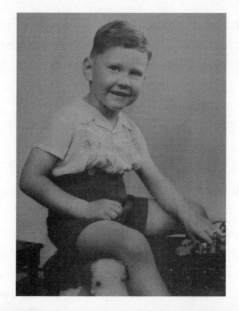

Top James Kilsyth Forrester, aged 3 years and 4 months.

Bottom Jim with his bagpipes at St Andrew's College, 1960.

Hurunui Division Commissioner and served in that capacity until 1978. She was a stalwart of the local Plunket movement, serving as Hawarden Branch President from 1968 to 1969, and as Vice-President of the Hurunui Branch in 1970. She was then elected to the Provincial Council and became its Vice President in 1982. Around the time I first met her, she was very active in a group called Friends of Waikari Hospital, and, with the fourth Labour government doing its best to close small provincial hospitals, Waikari needed all the friends it could get.

Ellen was a fanatical gardener and always had floral exhibits in the shows. This, of course, meant she was a member of Hawarden/Waikari Garden Club. And she was a staunch member of the Hawarden Branch of the Countrywomen's Institute, and a pillar of the Anglican Church, being a member of the St Columba Ladies Guild and keeper of the church linen.

And all the while, she was a mum. She was a keen sportswoman, being very good at golf and tennis, and a very good swimmer. Her children remember spending just about every summer weekend down at the river — after Mum had finished the ironing and gardening, of course!

JIM WENT TO PRIMARY SCHOOL at Hawarden, at the school where land was gifted for the purpose by his grandfather, James Forrester, whose picture hung in the staffroom until the mid-1990s. Jim — my Jim — was terribly upset when the school decided it should be returned to the family and gave it to me one day. After primary school, Jim boarded at St Andrew's College on Papanui Road in Christchurch. It was a boys' school in those days: it went co-educational in 1991, a development that Jim tut-tutted at when it happened, and that a lot of his old mates had trouble with. It was quite a strict school, as you'd expect of a place that was heavily influenced by Scottish Presbyterian values. But Jim did alright there, being a good rugby player and a very good member of the school's famous pipe band.

His mates don't remember him distinguishing himself academically, although he harboured ambitions of being a vet. One of his mates at school, Stewart Johns, wanted to be a detective in the CIB. Bruce Evans and Roddy Innes wanted to go flying. Funny these dreams that country kids have: you don't know how seriously they take them, because, of course, the reality is that they will work within the established family business of farming. Stewart went onto the family farm at Culverden. Roddy went onto the Innes place, and Bruce Evans had his wings partially clipped by expectations, too. And sure enough, as soon as he decently could, Jim left school to help his dad on the farm.

Frank was farming Heatherdale, and his son's help would have been welcome. In 1966, Jim bought Black Hills from the estate of his grandfather, Robert Adam Forrester, and in 1973 he bought back the Karaka Block, which the Cotterill family had bought from Will Forrester's brothers and sisters after Will's death in 1953.

By the time I came on the scene, Frank and Jim were running 3,000-plus Corriedale ewes (they also had NZ-registered Flock #52 stud prefix 'Heatherdale') and Jim had a hobby — 80-odd Herefords (stud name 'Blackhills', NZ Herd #690). Typical of a farmer to take up farming as a hobby! Lord knows what the locals thought when Jim's flat-deck truck rolled up with my coloured darlings aboard, and the next load contained my spinning wheels, weaving loom and a few bales of coloured wool!

Like the rest of the district, Frank and Ellen openly accepted me as a match for their Jim. A funny thing happens in rural

Top Beverley (centre) with her cousin, Brian Cole, and his wife Penny, who have been key members of the team getting BlackHills Yarns into the UK.

Bottom Three generations all belonging to Rural Women New Zealand. Noelene, Ona, Beverley, and Olive in front, taken in 2006.

Top Olive Phillips turns 100. Ona, Mansel, Noelene, Colwyn, Chris and Lynley, with Olive (Nan) and Beverley in front, 5 November 2007.

Bottom Jim and Beverley — a lovely photo taken while on the New Zealand Hereford tour just a few months before Jim died.

Top Beverley and Helen at the Christchurch A&P Show, around 1997.

Bottom Beverley's homestead in the snow.

Top Visitors from the
Shetland Islands to Black
Hills in 2014. We are
all wearing authentic
Shetland Isles garments.

Bottom left A Maasai
woman rolls BlackHills
fleece wool into yarn
on her leg.

Bottom right Teaching
Western knitting to
Maasai women in
Arusha, Tanzania.

Above Branded and colour coded sheep on common ground in the French Alps. The shifting of sheep from plains to highlands is called transhumance.

Left The ever-dependable Russell.

Top Beverley with Stewart Johns outside BlackHills Yarns wool shop in Henley-on-Thames.

Bottom Jude, Betty of Country Woollens and Beverley at the Christchurch show in 2013, with their ram that won Supreme Black and Coloured Sheep and third place All Breeds Wool.

BMW 3-Series Touring F31 (2012)

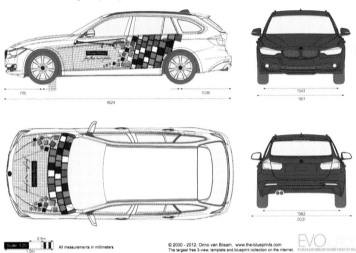

The plan for the decal wrap to go on a beautiful 3-series BMW, the garment we modelled our wrap on, and the car with the wrap in place. The Beverley Riverina Collection garment was runner-up in the BMW New Generation Award 2013 New Zealand Fashion Week.

New Zealand Fashion
Week 2013 — the opening
five-piece outfit (top)
and the closing garment
(bottom) from Beverley
Riverina Handknits.

families. As a woman, when you marry into one and live on the property, you pretty much assume the role of another daughter. Frank and Ellen's girls had all married and moved away: Rosemary and Wendy had both married in 1972, and Helen had married in 1977. By 1986, Frank was keeping pretty indifferent health, so I was regarded as a welcome addition to the family, maybe because of my experience with invalids and health issues.

I found myself balancing the duties of keeping house for Jim and my own occupational therapy practice — I was covering North Canterbury by now, just as everyone had hoped, but also increasingly ran Frank to and from doctors' appointments and did groceries and errands for both him and Ellen. They phoned every day to have a chat and tell me about things going on in their household, just as if I was their daughter. It was wonderful; I never minded a bit. I had been brought up in a community where this state of affairs was the norm — my own grandfather, Sam Price, was part of the family, living in our family home for eight years — so I suppose I took it all in my stride.

Funny thing was, after I got engaged to Jim, I asked them what they would like me to call them. I knew Jim's sisters called them 'grandma' and 'grandpa', because they had children. That couldn't work for me. But they invited me to call them Frank and Ellen, which raised eyebrows around the district a bit. I would introduce Frank to someone, saying: 'I'm Beverley, and this is my father-in-law, Frank.'

The person would look at me strangely.

'Yes, I know Mr Forrester,' they would say pointedly, as though I had been rude to royalty.

Frank died on 24 October 1987. I'm so pleased I got to know him. Like his son, he was a lovely man. On the Friday just before he died, Frank looked on as Jim, Russell Hassall and I were tailing my lambs down in the shed. He clearly wasn't well, as he eventually admitted to me.

I stopped what I was doing and looked him in the eye. 'You get yourself to the doctor,' I told him.

Later that afternoon, I was down at the garage at Waikari seeing about my car, and I saw Frank making his way to the medical centre. He didn't look too good, yet it was a surprise when the phone went the next day and Ellen told us that he had been taken off in an ambulance overnight. That's typical of farmers. They manage to hang on until night-time, as if they can't bear to leave their land in the daylight, as if they can't bear to say goodbye, or when they cannot see all that they're leaving undone.

We took Ellen down to the hospital. Frank was lying there, and I saw right away that he was in a bad way. Ellen and Jim were just standing there diffidently.

'I don't think your dad's too good,' I told Jim quietly.

'What do you mean?' he asked.

I organised a chair for each of them at Frank's bedside, and encouraged Jim to hold one of his hands as Ellen held the other. Very shortly afterwards, Frank's eyes opened and fluttered. Then he relaxed.

A nurse came and declared: 'Yes, he's died.' And then I fainted. I'm funny like that. I was fine while I was able to organise everyone else, but as soon as there was nothing to do, I got the wobbles. I went down in a heap, clipping my eyebrow on the foot of the bed on the way. I came to, and then was taken to the next room to have my eye stitched up. The stitches and the bruising made for an interesting line of conversation at Frank's funeral!

POOR ELLEN. THIS WAS A terrible time for her. She had lost an infant daughter and now she had lost her husband. And we soon got a sense of what the division of labour had been between her and Frank when she had to pay for something but had no idea how to set about writing a cheque. Jim then did it for her.

Of course, this was a difficult time for everyone. But it was especially hard for me, being a bit of a johnny-come-lately to the family, and married to the only son and oldest child. The girls arrived, and, while I had taken on duties in their absence, they were now home. This was a new situation for everyone. I am forever grateful for their help.

It became a real insight into the way rural communities organise themselves, and after that I often found myself giving talks about how, with the advances in medical technology, as farmers and their wives were getting older, often they would find themselves alone and rattling around in a great big house with all their children having left home.

The burden of running Heatherdale, which he had shared with Frank, now fell full-square upon Jim, on top of running his own properties. He worked both farms for five years or so until it started getting too much and it became plain he would need someone to live on the farm and help with the livestock. That meant freeing up a house, and, of course, the logical move was for us to move into Heatherdale and a livestock manager to shift into our house on Karaka. That's what happened. Ellen went to live in Waikari. It was her idea, but, of course, it was still a huge thing for her, leaving the home she had lived in and lavished such care upon and taken such pride in for the better part of 50 years.

>———▶

THESE WERE HARD TIMES ALL OVER. We had some bad luck with the weather. There were a couple of big droughts, which meant you had to send just about all your young stock off the place by December, and drafting was a horrible business in the sheds over at Karaka, with the wind plastering you with dust inside and out. I'll never forget one terrible snowstorm when Jim was away up in Feilding at a Hereford Breeders' meeting. Jim had left the crawler tractor up the back of Black Hills. He heard on the radio how bad the weather was getting and phoned, and talked me through the process of starting the blimmin thing.

'She won't start in the cold, someone will need to take a tin of ether and give 'her a good squirt in the breather. Got that? And don't forget to warm up the glow-plugs, will you?'

'No.'

'Good on you.'

We got the machine going and managed the hills, creating a track for the cows to follow — and, incidentally, for Telecom and TV3 to get access to their repeaters on top of Mount Alexander.

Another time when the snow finally melted, Jim was terribly upset to find groups of four or five calves dead in huddles. What cows will do in bad weather is leave a nurse animal in charge of a group of calves while they head off to find feed. Once they've fed, they'll come back and retrieve their little ones and the nurse is off-duty. But, of course, if the snow gets too bad, they never get back to their calves. It's at times like this you're supposed to shrug and remind yourself that this is nature's way, and that where there are livestock, there are also dead stock. But Jim never found it that easy, and nor did I.

Nor were the run-of-the-mill hardships of farming all we had to contend with. By the end of the 1960s, New Zealand's charmed run was all but at an end, and there was talk of Great Britain entering the European Economic Community (EEC).

180

Up to that point, Britain had not only been our most loyal customer, but it had also been our champion at the European table, ensuring our produce could enter European markets on favourable terms. If Britain were to enter the EEC, we would lose both, and we would be at the mercy of European trade terms which, with their own subsidised farming lobbies calling the shots, were always going to be pretty mean.

So it proved. Britain entered the EEC in 1973, and while there was a system of quotas in place to ensure we could find a market for a substantial quantity of our meat, wool and dairy, it was substantially less than we were geared to produce. Rather than let nature take its course and allow farmers to face the full horror of market forces, the New Zealand government chose to extend all kinds of subsidies to farmers to prop them up: fertiliser was subsidised; loans were available at special, advantageous rates of interest; there were tax breaks and even financial assistance with overheads such as pest control, notably weeds and rabbits. New Zealanders in the 1970s lived in the shadow of growing 'mountains' of butter and meat.

The most blatant measure of all was the 'supplementary minimum prices' scheme brought in by the budget of 1978, supposedly to maintain the confidence of the farming sector in its future prospects. After all, the government pointed out, New Zealand agricultural produce made up fully 70 per cent of our export earnings, with forest products accounting for much of the rest. But in reality it was all a bit of an election bribe, and it meant that, as the 1980s came around, we were living in a fool's paradise. Jim and Frank had much the same attitude as most Kiwi farmers: they were happy to take the breaks while the government was prepared to extend them. The economics were the government's business — farmers just got on and farmed.

It all came crashing down in 1984, when David Lange's fourth Labour government swept to power. Its first act was to devalue and then float the New Zealand dollar. Interest rates rose and took the dollar with it, which meant that export earnings plunged. Meanwhile, the government didn't muck around in dismantling the whole apparatus of subsidies, too: farmers soon found themselves paying market rates for finance and fertiliser, and by the end of the 1980s they were on their own when it came to tackling nasties such as rabbits and nassella tussock, too. This was a terribly hard time. The stress was unbelievable.

AS WITH THE WEATHER, there's not a lot you can do about periods like this, when the political pendulum swings and you cop it. North Canterbury had experienced a bad drought throughout 1982 and 1983; the sudden change in the economic climate was just another one of those things. You put your head down and got on with it.

At least with the hydrological cycle, there were steps you could take. Since the late 1970s there had been plenty of enthusiasm for community water schemes, and Frank, Jim, Russell and all the other men in the district had installed miles of pipes to bring water from the Waitohi River to tanks on individual properties. Now it was possible to put troughs on the farm, where previously a dried-up creek had meant you had to de-stock, plain and simple. Whereas previously the maximum carrying capacity of your land had been defined by what you could hold in the driest part of summer, with water available year-round it went up.

But one of the unexpected effects of the withdrawal of government subsidies and support for farmers was the spectacular comeback of an old farming foe: those pesky rabbits. With subsidies for pest destruction removed, rabbit control became yet another overhead for farmers who already had their backs to the wall. The inevitable consequence was that it just didn't get done, and the little blighters started overrunning the district. North Canterbury was as hard hit as anywhere. At times it looked as though whole hillsides were moving, such were the numbers of rabbits. The grass was eaten down to the dirt in places, and sheep and cattle were all but starving. Jim and I saw grown men reduced to tears by the damage done, and quite a few were ruined.

We were spending $20,000 a year on rabbit control; we knew people who were spending four times that. Jim seemed to take it all personally, as though it was his stockmanship or some other aspect of his farm management that was to blame. Like many farmers — maybe even most farmers — he had a tendency to do that. I remember how badly he took it when the herd tester told him that one of his cows was positive for tuberculosis. Jim had taken all prudent steps to protect his herd from TB, but you can't control where the vectors (most notably possums and ferrets) pop up. It's the luck of the draw, really. Jim knew this, but it didn't stop him feeling responsible and bitter, because having a single reactor in his herd meant he was banned from taking his beloved Herefords to the shows.

Rabbits were introduced to New Zealand at several points around the country in the nineteenth century, supposedly to form the basis of a local fur trade; but the local invasion began up in Kaikoura, where Captain Kean of Swincombe Station liberated a dozen silver-grey rabbits in the 1850s. Within half a dozen years, his property was so overrun that he had to walk off. The spread from there was remorseless.

As rabbits became a nuisance south of the Waiau, runholders combined to build a 135-kilometre rabbit-proof fence from the sea to the Great Divide, in 1888. This held for a few years, but once it fell into disrepair the advance resumed apace. Rabbit

Nuisance Acts were passed by the government in 1876 and 1882, and Rabbit Boards were established in most districts, including in the Hurunui. Desperate remedies were tried: in 1888 the Amuri Rabbit Board imported 1,000 ferrets, 1,000 stoats and 400 cats, and liberated them in the district. It topped up the populations the following year, with a further 600 ferrets and stoats and another 400 cats. All the same, in 1894 a report to the Hurunui Rabbit Board noted just how bad rabbit numbers on Glenmark had become. As it turned out, this was the height of a plague. The population probably overshot itself and settled back again until the 1920s, when there was another spike. This continued to be the pattern for the next 50 years.

For most of the century since Canterbury was settled, there was a bounty on rabbits, and most runs had a rabbiter or two who were kept fully occupied shooting and trapping. By 1893, over 17 million rabbit skins had been exported to England, and 40 years later this number had climbed to 24 million. But there was yet another population explosion in the 1940s — probably because all of the rabbiters went to war, or took up the farm jobs that the young men left open when they enlisted — and the government began casting around for new and more effective methods of fighting them.

It passed a Rabbit Nuisance Amendment Act in 1947, which directed all Rabbit Boards to concentrate on killing rabbits, rather than merely controlling or even trying to harvest them for commercial purposes. All manner of poisons and baits and combinations thereof had been tried down the years, with mixed success. In 1947, it was shown that the best and most effective was carrots laced with sodium monofluoroacetate — that's 1080 to you and me — and spread by plane, which became the weapon of choice in the 1950s.

None of it worked for long. The pattern was that the population would seem stable until there was a drought, whereupon rabbits seemed to hit their straps. It was a combination of the drought of 1983 and 1984 and the progressive reduction of government control measures that saw the rabbit population in the Hurunui spiralling out of control again. People began looking around for solutions again. One that looked promising was biological control — disease.

The government had tried myxomatosis in New Zealand in the 1950s, with little success. This virus is a scourge of European rabbit populations, and the New Zealand government presumed it was just a matter of letting mxyomatosis go, a bit like the way you'd light the touch-paper on a skyrocket, stand back and watch. But myxo was a fizzer in New Zealand because New Zealand rabbits aren't infested by the flea which spreads it amongst its European cousins, and so, in the absence of any other suitable vector, infections remained isolated. Still, there was a considerable lobby for trying again, and this time doing it more carefully.

Meanwhile, something was killing rabbits in Asia. Around 1983, most of the rabbits in China started showing symptoms of some kind of haemorrhagic disease, a bit like a bunny version of the awful ebola disease. The mortality rate was spectacular. In 1984, a Chinese biologist reported that a mutant version of the rabbit calicivirus that was endemic to European rabbits seemed to be the culprit. Soon afterwards, the Asian strain appeared in populations in Europe with the same devastating effects. Well, New Zealand and Australian farmers knocked on the government's door and said: 'We want what they're having.'

In 1991, the Australian government imported the virus from Europe with a view to testing it under quarantine to see if it could be safely used as a biological control measure there. The New Zealand government was watching over their shoulder. The site chosen was Wardang Island, off the coast of Yorke Peninsula in South Australia. The trials were promising: Australia's indigenous fauna seemed to be unaffected by rabbit calicivirus disease (RCD), as it was then known, but it was brutal on rabbits. It was clearly good enough for someone, because the virus seems to have been smuggled to the Australian mainland, and within a couple of months of its first appearance it had killed nearly 10 million rabbits.

Still the Kiwi government dithered. There didn't seem to be any good reason for their caution, given that the effects of the virus in Australia seemed to be confined to the rabbits. There didn't seem to be any risk to stock or people, and the chances that a mammal virus would pose a threat to our native birdlife were pretty remote. The government looked into it, and we even had a visit from a party of seven vets led by a Ministry of Agriculture official, Peter O'Hara. Jim showed them over the more devastated parts of the district, and got them to talk to the young family next door, whose livelihood was under direct threat from the rabbit plague. But after studying and studying the matter, the government announced on 2 July 1997 that it would not allow RCD to be imported and released. I remember hearing the announcement over a radio that was playing while I was at a patient's home. I was furious, and so was Jim.

Within a couple of months, though, rumours had it that a bunch of farmers down south had gotten hold of the virus and released it. As Chairman of the Pest Board, Jim was regarded as both a leader and someone whose judgement in the matter could be trusted. I'm sure the wisdom of releasing the virus was discussed. There were plenty of theories around about how it might have come to happen. Someone working on the effects of RCD on the physiology of rabbits might have told someone else that all you would need to do is take tissue from the liver and spleen of a rabbit that had died of the virus. Expose other rabbits to this, and infection of live populations could swiftly follow. Perhaps someone visited an RCD-infected farm in Australia with a sample jar and told a few porkies to the border control people on their return to New Zealand.

Whatever the truth was, farmers could now take matters into their own hands.

In September it was revealed that Otago rabbits were dying of RCD. Information on how to go about infecting your rabbits with the virus spread just as fast as the virus itself, and some people were reportedly making 'rabbit smoothies', which were then spread around rabbit burrows or added to carrots and scattered in paddocks. It was a calculated act of civil disobedience: the authorities were keen to find out who was responsible for the importation and spread of RCD, but no one knew. Farmers were just as keen as police and the Ministry of Agriculture to find the bloke, because I don't know of any farmer who wouldn't have shaken his hand and bought him a beer. To this day, no one has been prosecuted and no one knows whodunnit. But by the end of 1997, we were picking up dead rabbits all over the property, the grass was coming back, and you could just about hear the stock gaining weight again. If only Jim had lived to see that.

JIM AND I CELEBRATED our tenth wedding anniversary in late 1996 by going overseas. He took quite a bit of persuading — now that I'm in his role, I know how hard it is to ignore the call of the million and one things there are to attend to on the farm — but he agreed to go, and we visited England and went to the Chelsea Flower Show. Afterwards, we did a Trafalgar tour of Europe. It was Jim's first trip further afield than Australia. He eyed the antiquities and monuments of Europe from the tour bus with a critical eye. He was impressed — I could tell he was — but he didn't show much excitement. I do remember one day when we got off the bus for our morning-

Jim and Beverley.

tea stop in some lovely old town square in Belgium, and he looked at the old cobbles at his feet and gave a start. 'Good God,' he said. 'They've got blimmin barleygrass here, too!'

We got to Venice, and I had told him that part of doing package tours was doing all the add-on trips that cropped up. We were being given the chance to do a gondola ride, and Jim had thought that sounded alright. But when he realised the gondola was a boat, not a cable car, he shook his head, crossed his arms and said: 'I'm not going up all the sewer traps of Venice.' He did go, in the end, and he enjoyed it. He even thought the piano accordion player was better than me! It goes to show you can take the farmer out of North Canterbury, but you can't take North Canterbury out of the farmer.

Away from the place, we had the chance to take stock of our situation back at home. These had been hard years, but we looked like we would get through, especially if the rabbits could be knocked back. There might well be much better times ahead. No one was getting rich out of farming in the 1990s. Wool prices had fallen steadily throughout the 1970s and 1980s as the world's consumers became infatuated with synthetic fibres. And having been spoiled by preferential access to British lamb lovers for most of our history, little thought had been given to how to market sheep meat to people who weren't used to it. Consequently, from the 1970s onwards, the bottom fell out of meat, too. Add to that the upheavals of deregulation, and you can see why farmers were doing it harder in the late 1980s than they had been doing it for half a century.

Diversification was supposed to be one answer. Across New Zealand, all kinds of weird and wonderful plants and animals were being tried on properties that had once grazed sheep and cattle: deer, ostriches, alpacas, angora goats; even — God help us — rabbits. Kiwifruit had been a big success up north, and so people tried babacos, olives, macadamia nuts and so on. Lots of marginal land had gone under pines, as there was a boom in forest products to Asia. And wine was a biggie. Like most of the locals, Jim was pretty sceptical when John McCaskey stocked some of his land with Santa Gertrudis cattle and planted a few blocks out in grapes. But within 10 years, everyone was eating their words: some Christchurch doctors had founded Pegasus Wines just down the road at Waipara, and the vineyard was flourishing, just as the stock of New Zealand wines was rising overseas and amongst the new breed of person you saw driving flash cars around in Auckland.

Our land has potential. In places, it has the same deep, gravelly soils and terraces over a limestone base that winegrowers look for. But Jim was a traditionalist, a sheep and cattle man through and through. In those days, we were running around 3,500 Corriedales and 70 or 80 Hereford, breeding across the farms. Jim did most of the management himself, ably assisted by Kevin Hislop and then Bob O'Carroll on Black Hills and Karaka. Don Rowe was casual, and Russell Hassall was full-time on Heatherdale.

And I had my hobby, as Jim slightly conscendingly called it. I had built my flock of coloureds up to around 150 head, and they kept me happily occupied. When Jim took his cattle to the shows, I took my sheep, and we both did pretty well with them, too. We both had our commitments, judging and so on. With Jim's encouragement, I still had my part-time job as North Canterbury District Occupational Therapist, which got me around the countryside a fair bit, from Amberley right up to Kaikoura, and helped me to get to know, and get known by, pretty much everyone as Beverley, not just as Jim's wife. It was an entrée to the farming community; but, by the same token, my connection to a local farm was very useful, too, in building a rapport with my rural patients.

I had dropped in to visit one family who had lately arrived from the North Island to set up in the Hurunui and who needed my services. I was waiting in the hall downstairs as my patient made his painful way down. I happened to notice the Hereford Breeders' Association magazine on the table, and this was an instant point of connection.

'I see you're interested in Herefords,' I said, as soon as we'd made introductions.

'Yeah, we run a few.'

'So do we.'

'Oh, you're that Forrester, are you? Down the road there with the Herefords on Black Hills?'

And that kicked off a conversation about farming and cattle breeding that broke the ice, so that when it came time to turn to the reason for my visit we already had a rapport. He'd had a stroke, and his recovery had been slow. My job was to find out what problems he was having getting on with life and to find solutions.

'You know, Beverley,' he said sadly, 'my biggest problem is that my daughter's getting married in six weeks, and I want to walk her down the aisle.'

'Right,' I said, 'forget about everything else — that's what we'll get you doing.'

I remembered that Princess Diana's wheelchair-bound father had walked her down the aisle by leaning on his daughter — a case of who's walking who — and that's what we decided we'd do. Once I'd got him to stand and even walk a few unsteady steps, I showed his wife how to support him, with a view to her passing the technique on to her daughter when they travelled north for the wedding.

It wasn't an easy process, and there was lots of frustration and a few tears along the way. But we talked it all through, and, whenever we needed to, we turned the conversation to farming. One evening a few days after the wedding, I got an emotional phone call.

'I did it, Beverley. We made it all the way to the altar.'

Who knew the Hereford Breeders' Association magazine could do so much good in the world?

I WAS HAPPY! After the business with the previous divorce, it had taken a while to get back on track. Work and coloured sheep had helped, and then meeting and marrying Jim, of course. He had restored my faith in men. The only real regret I had at that stage was that we had no children. Jim would have loved to, and so would I. I think I had always sort of expected that I would, but it was not to be. I think I was probably sadder for Jim. But this aside, I was happy.

Russell Hassall works with
Beverley on the farm.

8

ALONE AGAIN

THERE'S A GOOD CROWD in the Hurunui Hotel tonight. The occasion is a 'snow shout', which is supposed to be the community's way of saying thank you to the people who have been working the machines and plying the rakes and shovels that cleared roads and driveways after the big dump of snow we had in September. But really it's just another excuse for a get-together. That might sound a bit gratuitous, but the reality is that farmers can get pretty isolated. I can spend three or four days on the farm without seeing another soul, longer if I don't make an effort to get off. We're all the same. So I have always believed it's important that people make the effort to get together. I always encouraged Jim to drop into the Hurunui for a drink or two with his mates at the end of the working week. It was important he got that kind of outlet.

Just how important is brought home to us tonight. Doc Sidey speaks, and speaks thoughtfully, about exactly that:

isolation. Communications have improved since the Hurunui was settled — roads, bridges, telephone, internet (when they're working). But in some ways, the ease with which we can stay in touch has actually broken down some of our sense of community. Instead of shopping or entertaining themselves locally, people are just as likely to hop in the car and head to Rangiora, or even to Christchurch.

There are few other occupations where your life is bound up so tightly with your work than farming. Mostly, farmers live in their workplace. They're on-call 24 hours, 7 days a week, particularly when the weather is threatening, or lambs are arriving. Flukes of nature or world economics can wipe out in an instant a couple of years', if not a lifetime's, hard work. That farmers are accustomed to take the rough with the smooth is illustrated by that saying 'Where there are livestock, there are dead stock.' You're closer to the brutal realities than most. It can be pretty hard to get your head out of that space. Some don't manage it, and it can take its toll. Sometimes the toll is terrible. Doc is speaking about the son of a nephew of his, just a young fellow, who was farming until he took his own life recently. Typically of the Hurunui, I knew him well, and his dad, Jim Sidey, even worked for me on Karaka for a while.

WHATEVER PERIOD OF HISTORY YOU READ, farming the Hurunui has been hard and you'll find tales of heartbreak. Back in the days before the rivers were bridged, the isolation was real and physical, especially when it rained and the rivers swelled. You could spend weeks stranded between the Waiau and the Hurunui to the north, and the Waipara and the Waimakariri to the south. And even when the rivers

Top Hurunui Hotel   **Bottom** The old rabbit-proof fence, as it exists today. The cost of installation in 1863 was £50 per mile.

were crossable, the limitations of the modes of transport available to you made distances that seem insignificant today just about impossible back then. If every time you wanted to go to town you needed to catch a cussed horse and tack it up or hitch it to your gig, never mind enduring the rough, bumpy and dangerous ride itself, I'm sure it would make staying put the more attractive option.

In 2012, I bought myself a 1937 Austin Big Seven. It's pretty basic compared with modern cars, and most people would think it's a dreadful old thing to drive. I love it. And you can see what a marvellous invention the motorcar must have been for women at the best of times, and what a lifesaver it was for women expecting children or nursing sick family members. In her book *Hurunui: Source to Sea* — a love-letter to the Hurunui, really — Shona McRae writes of a spot under some trees close by where the Glen Ghon Stream joins the Kaiwara Stream that serves as a constant reminder to her of how hard the lives of our pioneer forbears were. Around 1859, the Cheviot Hills boundary rider and his wife set off on foot in bad weather from their isolated home near the Kaiwara to try to get their little boy to a doctor. He died only a few miles from home, and there was nothing for it but to bury him in that spot.

That's one example of the effects of isolation. Another is the pitiful story of Kate Lohenet, the wife of Alexander, the cook at Glenmark. (His surname was variously spelled in the newspapers as Hobenet or Hobinet.) On Saturday, 25 October 1896, she went for a walk with her six-year-old daughter. At some point, she sat sadly on the low wall beside the fountain in front of George Henry Moore's grand house, kissed her little girl and told her to go and find her father where he was out on the farm, give him a bottle of 'herbal beer' and tell him goodbye from her. When the child got back, she found her mother floating in the pond where she had drowned herself.

This and countless other stories just as sad were heard

down the years by a coroner sitting right here at the Hurunui Hotel. It's a place of good cheer these days, and likely always was, but in the pioneer days there was a room there set aside as a morgue, and it was common for the coroner to hold inquests in the hotel. Of course, the rivers often made it impossible to get the bodies of the deceased out to the wider world — and all too often the rivers were the culprit in the tragedy. So it's quite easy to believe the popular notion that the hotel is haunted.

IT'S THE ISOLATION, PHYSICAL AND MENTAL, of farmers that makes community so important. That's why there is such a festive atmosphere at regular events such as the A&P Shows and the sales, the dog trials and the Hurunui races: it's more than just getting together with your friends and neighbours and having a good time. These events are an affirmation of community, that you're all in the same boat: a chance to satisfy yourself that all is well with them, and to show them that all is well with you.

Some people find the closeness of rural communities overbearing — the nosiness, the constant scrutiny. But it's reassuring, too. You can be driving past a neighbour's place and see something out of sorts — stock in a paddock without water, a fence down or wandering sheep or cattle — and you immediately pick up the phone or drop in to check if everything's okay. When you call on neighbours, you stay for a cuppa and a chat, because that gives you more time to catch up and get a feel for how things really are. And then if I need something, down the track, the time I have given makes me feel as though I can ask.

And you find that the community gives back in pretty much

equal measure to what you put in. If someone rings me and invites me to an event I can't attend, instead of just saying 'No, thanks, I can't' or something like that, I'll add: 'But please ask me again.' Otherwise, people will try once or twice and then decide you're not interested and give up. It's little things like that that keep up your ties.

More often than not, you'll find farmers are pillars of their community — they'll not only show their stock and compete at dog trials and play rugby or cricket, but they'll sit on the committee of their local sports clubs or A&P association or Road or Rabbit Board and so on. They'll be members of the local volunteer fire brigade or RSA, and they'll be avid churchgoers, even though they're not especially religious.

Jim was no one's idea of an extrovert, but he ticked most of these boxes. He was on the Rabbit Board, and he was made a life member of the Canterbury Country Rugby Union. He was on the committee of the Canterbury A&P Association. He was deeply involved in the efforts that were made to rescue and restore the Hurunui Hotel when it came up for sale in 1979. And he was on the receiving end of heaps of pressure to make himself available for election as president of several organisations, but he always declined, on the grounds that he wasn't married.

In a deeply traditional community like the Hurunui, these roles are a two-person job, and Jim didn't have a wife to discharge the duties of the president's 'handbag'. Believe me, there are a hundred and one things to do, as I discovered after I married Jim and he could no longer use his traditional excuse to wriggle out, and suddenly became president of the Hawarden A&P Show and the Waikari Dog Trials. He was also in demand as a piper, and used to play his bagpipes at Hawarden ANZAC Day services years ago, until he broke a finger in the cattleyards one day and could no longer do it.

I'm a member of Rural Women New Zealand (formerly the

Women's Division of Federated Farmers), like Mum and Nan were — I think we were the only family with three generations as current, active members at one time. I'm on the committee of the Canterbury and the Hawarden shows; I regularly show my sheep and serve as a judge all over the place. I'm in the Anglican Church and I attend periodically. I also attend NZ Creative Fibre and local woolcraft days as often as possible. Like both my parents, I'm a member of the RSA, and I think I've only ever missed one ANZAC Day service in all my 63 years. And, as I explained earlier, my job helped cement my place in the community, too, with an identity that was separate from Jim's.

These are the kinds of things you do, because you have a stake in your community. You have to put the effort into your community to make sure it's there for those who need it. And most of us need it, sooner or later.

ON FRIDAY, 12 SEPTEMBER 1997, Jim had gone off up to Culverden to get some seed from Bob Frame's place and to visit George Poppelwell, the engineer. He'd had a pretty big day. We were TB testing all morning, and he was crutching two-tooth rams all afternoon, and on the way home he'd dropped some gear off at the gateway of an OT patient of mine. I had encouraged him to call in at the Hurunui on the way home, as he did most Fridays. That's what he did, and he arrived home with the fish and chips for our dinner around six in the evening as always. He'd had a beer with his mates down at the pub, but not too much: he liked a few whiskeys every now and then, but he was never a big drinker.

We had settled down in front of the telly to watch Maggie Barry's *Garden Show*, and it featured the Chelsea Flower Show.

'Look!' he said. 'We've been there! We were there! You got me there, didn't you?'

'Kicking and screaming,' I said.

That made us both giggle, remembering how hard it had been getting him to agree to go — typical farmer, you virtually had to chain him to the tractor and tow him out of the gate. He laughed. I was knitting, as usual, and I got up to close the curtains. I asked him where he'd like to go next, and he was halfway through answering when he stopped. I turned around and saw that Jim — my lovely Jim — had gone, no fuss, just like that. I got him on the floor and tried CPR, but it didn't do any good. I picked up the phone. I didn't know who I was supposed to ring, so I phoned 111 and then my neighbours.

It's a bit of a blur. The police and my neighbours arrived, and it dawned on me as I tried to answer their questions that the police were treating Jim's death as suspicious, and me as a possible suspect. It took a while before they seemed to relax a bit, and soon they were kindness itself. More people arrived — Jenny, my friend also from Victim Support, and Richard, my GP. It was a surreal time. And of course if Jim's sudden passing was a shock to me, it was 10 times worse for poor Ellen, who would never have expected to survive her son. Jim was only a whisker shy of his fifty-third birthday.

THERE WERE THINGS TO BE GRATEFUL FOR. I was right there. We were in the middle of a lovely moment, laughing about one of our best times together. He could have died in the bath. He could have had a heart attack at the wheel and hit the abutment of the bridge over the Hurunui and left me wondering about the hows and the whys and the whats. He could have been alone, but instead I was right there. I've been there at many of the deaths in my family, and others, and I wouldn't have it any other way.

Word got around pretty quickly, and as I drifted through the weekend in a bit of a daze, the phone kept ringing and people kept dropping in with frozen meals and flowers and offers of help. There were a few people about the district who knew about the whole divorce business, and they were aware that this was the second time I had suffered a loss. That seems to have made them extra solicitous of me. The kindness was humbling, and it got me through the worst time, the couple of days immediately after Jim's death, days spent with disbelief and the sudden, awful realisation repeatedly washing over me like the Pegasus Bay swell.

On Monday morning, bright and early, there was a knock on the door. A bloke was standing there, with a digger on a truck behind him.

'Morning,' he said. 'Jim about? I've come to do those dams.'

It took me a while to recall that Jim had said something about installing some stockwater dams up in the Black Hills. I suddenly became conscious of the whole weight of the running of the farm, the burden that Jim had borne so lightly on those broad, capable shoulders of his, settling on mine.

'Look, sorry,' I said, flustered. 'I'm not really sure where he was going to put them. Jim's gone, you see.'

'Oh, righto. When's he back?'

'He died. On Friday.'

He stared at me.

'Oh, bugger me,' he said. 'Jeez, I'm really sorry.'

All I could do was nod.

The digger driver and I decided he would have a look up in the hills and work out where he thought the dams should go. That was one problem solved.

ONE DAY, A BEAUTIFUL FINE day just before the funeral, I packed a thermos and some nibbles and put Ellen in the car and we set off up the hill towards the trees around 400 metres behind the house.

'Where are we going?' she asked me.

'Oh, we'll just nip up the back paddock and have a picnic,' I said.

She smiled strangely.

'Do you know, I've never been up this far before?' she said.

Three weeks after the funeral, I was obliged to attend a Hospital in-service course. It was CPR. I asked quietly if I might be excused: sometimes it just doesn't work.

What to say about Jim? He was a quiet bloke, as farmers often are. But like a lot of farmers, there was a lot going on under the surface. People trusted Jim, and whenever there was a bit of a dispute about an issue, both sides would bring it to him. Because the thing with Jim is he never rushed to judge. He'd always hear people out on every side of an issue before he made his own mind up, and he'd take his time thinking it over, too. He was the soul of sincerity. He took all his responsibilities seriously. He cared about his land, his animals, his family, his friends and, of course, about me. If he took on a role, he would discharge all the duties it entailed scrupulously.

Not that he was dour. He had a wicked sense of humour, and you'd see it from time to time, usually when he was comfortable in the company he was in, and it helped if he had a good whiskey in his hand. His mates — Doc Sidey, Stewart Johns, Ross Johnson and the rest of those St Andrew's boys, and his rugby mates — tell lots of stories about him and his exploits in younger, more carefree days: such as the time they went to a Hereford gathering in Australia and left Jim behind there. When he came back a few days later, he'd just smile quietly and shake his head whenever he was asked what he'd been up to. The time they tried coaxing his Austin Tasman up Baldwin Street, the steepest street in New Zealand, when they were in Dunedin, probably for rugby. They got most of the way up before gravity prevailed, whereupon it took a bit of swift manoeuvring as they gathered speed in reverse to poke the rear of the car into someone's driveway so that they could go down frontwards. The time they all went back to Jim's after a night at the pub and carried on drinking whiskey. Jim was gone for a little while, and meanwhile his mates spotted a large turkey in the yard which looked like dinner to them. When Jim came back, they had an attack of the guilts, and told him what they had done.

'Oh, I wouldn't worry,' he said, and then allowed a look of faint alarm to stray across his face. 'As long as it wasn't Dad's *prize* turkey . . .'

All those times.

I REALISED THAT PROBLEM SOLVING was what would get me through: looking at the situation, identifying the problems and solving them, in exactly the way I had been trained to do in my work in occupational therapy, and in the way my parents had taught me by word and example from the moment I was born.

Everyone was watching me to see what I would do. Not in an unkind way, but you could tell that they were all wondering whether I would carry on, or whether I would just pull the pin and bugger off. I suppose there were plenty of moments in which I was wondering the same thing. But at some point, pretty early on, I must have decided I would just get on with it. Because that's what I did. I carried on with the farm.

Everything seemed hard. From the moment Jim died, the bank accounts were frozen, of course. The trustees and I had to negotiate with the banks to make sure there was an account with money in it available to me so that I could pay bills until the will was probated. And while Jim had never set out to keep me in the dark about the farms' operations, information had come to me on a strictly need-to-know basis, and I realised right away how little I knew about the day-to-day running of things.

I got Jim's workers to come and meet with me. I didn't want to do this at my kitchen table: that seemed too personal. So I cleared off the kauri table in the woolshed and called them up one by one. I still remember the look on Russell's face when I asked him to help me write down his job description for me. He'd been working on Heatherdale for over 40 years: he'd never heard of a 'job description', let alone read or written one. But I had literally no idea what he did from day to day, and I needed him to sit down with me and talk me through it. I did that with the others, too, with Bob O'Carroll, Peter Mallinson and Don Rowe. I got them to tell me what they did, and what they would need me, their new boss, to do for them. I told them that the accountant and I needed to know their jobs. The conversations were pretty stilted.

'So, Don. What days and hours do you work?'

'Oh, just whenever, really. Mostly around cattle mustering and ewe fairs and stuff.'

'So what time of year is that?'

'Oh, reckon when the cattle need mustering, and the ewes go to the fair. About then.'

'How many hours does it take?'

'Oh, a few days. Sometimes it takes a bit longer. And sometimes Jim had me do other stuff, too.'

'What sort of stuff?'

'Oh, just whatever needed doing, really.'

'How many hours?'

'Just however long it took. Sometimes longer.'

'Once a week? Twice a week?'

'Yep. More, sometimes . . .'

I'm pretty sure they all expected me to chuck it in, but they never said so. They were all absolutely marvellous. They were respectful and as helpful as it was possible to be. For my part, I was aware they needed reassurance that their jobs would continue. I went out of my way to make sure they knew they were valued. With Jim dying, they had lost a workmate and friend, too. In many ways, I suppose it was just like being tossed in the deep end at Templeton Hospital all those years ago. The way forward seemed to be to get everyone working as a family.

One of the vexed issues was what — amongst all the plant and machinery — was owned by Jim's estate and what was owned by Frank's. I sat down with Russell and got a piece of A4 paper and a clipboard and drew a line down the middle and headed one column 'JK' and the other 'FK'.

'The truck. The Bedford. That was Jim's,' said Russell. I wrote it in the 'JK' column.

'The small baler. That was jointly owned. Frank and Jim went halves.'

I made another column: JK/FK.

I could tell Russell was beginning to get the hang of it.

'And you know the 10-ton roller? Half of that actually belongs to Paddy Bamford down the road. Jim went halves with him.'

A fourth column was called for: 'Other'.

People began phoning or showing up with farming equipment that Jim had loaned out all over the district. It was quite amazing, really: the honesty.

It all took so much sorting out. I couldn't have done it

without Russell, or for that matter without David Hastie, the accountant and, incidentally, a descendant of the Hasties who ran the Hurunui Pub in the late nineteenth century. Whenever I was tearing my hair out over some matter or another — the tractor would need fixing, for example, but, because it was jointly owned by the estates of Frank Forrester and Jim Forrester, the mechanic would be asking how to divide the cost of five bolts fifty-fifty — David would say, 'Leave it to me. Don't you worry about that stuff. And if anything like that arises, day or night, don't hesitate to give me a call even if it's 3.30am.'

As one of the executors of Jim's estate, Doc Sidey was amazing, and so was his co-executor, Ross Johnson — another of Jim's mates from boarding school, and best man at our wedding. Like Doc, he knew Jim's operation backwards, and he was adept at anticipating problems before they flared up. When I give talks to people these days about succession planning in rural businesses, this is one of the things I'm really emphatic about. You need to appoint executors who really know the business. People like Doc and Ross.

ONE DAY IN NOVEMBER I had a regional council weeds inspector show up and tell me that I had nassella tussock up in the Black Hills. Nassella is a nasty bit of work. It's native to South America, and arrived here sometime in the early part of the twentieth century. It loved conditions in New Zealand — and especially Canterbury — and began to spread rapidly. A single plant can produce over 100,000 seeds each season, and these are spread by wind, water, stock and farmers. Seeds can lie dormant for a decade. It's unpalatable to stock and, because it is taller and faster-growing, it out-competes palatable grasses.

By the mid-1940s, it had become such a problem for some farmers that they were forced to walk off their land, and the government recognised it as a major threat to agriculture with the Nassella Control Act 1946.

The best way to identify nassella is to fly over affected land in November, when it is flowering, as its purple flowers make it stand out from the tan and dun colours of native tussocks.

Well, this bloke was telling me it was up there; I couldn't muck around.

'You'd better show me what it looks like so I know what I'm dealing with,' I said.

He looked pointedly down at my elastic-sided boots.

'Well, you won't be going up there in those, will you?' he said.

I told him I would go wherever he needed me to go. He shrugged, and then led me a merry dance up and down the hills and over ridges and through creeks. It was blimmin hard work keeping up, but I wasn't going to let him know that. After two or three hours, as we stood up there on the skyline amongst the power pylons, he looked at me completely differently, as though he'd decided 'she can do it'.

Not that everything was easy. One day, I was climbing the front fence and I got my trousers hooked up. I tried to free them, but only made it worse, and, because it was a rabbit fence, it had a lateral strand of wire that made it just about impossible to lift my foot. I must have looked ridiculous! The only practical thing to do was to get out of my trousers, which I did in stages, pausing every now and then to act nonchalant as a car went past on State Highway 7.

Practical steps. That's how I dealt with everything that came my way. Still, if you'd told me at the beginning of the year where I would be at the end, I'd have laughed at you. I was a widow at 46, and I had become shepherd, cattle woman, cook, amateur mechanic, novice fencer, assistant solver of rural water-supply leakages, a learner tractor driver, team leader, bookkeeper, senior strategist in a rural enterprise, a thinker, a planner and a doer. Good job I was a problem-solver, because problems came at you every day in a blizzard. It took all my mental and emotional energy to work out which order to deal with them, let alone to solve them. There was no time for brooding, self-doubt or even self-pity. I just got on with it.

My team helped — Bob, Peter, Don and the wonderful Russell, along with any number of willing hands from around the district, where they were needed. My parents and sister were enormously supportive. I was lucky my dad and my mum were farmers and still farming: I used to phone and ask the most stupid questions. Ellen was wonderful, too. Nan would phone me to talk about the stockmarket (she was 90), and she'd check up on me, too.

'Have you got any bottom in your hay paddock?' she'd ask.

And all around me, the community was there, just as it was when Jim's grandmother, Bertha, was widowed at 53; as it was when Harriet Perrott found herself alone; as it doubtless was when Major O'Connell died and left Sarah O'Connell alone in charge of Mount Grey Station back in the pioneer days.

ONE OF THE THINGS that came to light when Jim's will was read was that he owned the 250-acre block of land that was used as the Black Hills hogget block, and that was known as 'Perrotts'. Jim had bought it in 1966 when he was 21 from the estate of R A Forrester. It now passed to me, which meant that I was now the owner of the ruined collection of historical buildings, including Stoneycroft, the little cottage where Frank and Harriet Perrott had lived, and where Robert and Louisa Forrester and their family had lived, too. No one is entirely sure when the cottage was begun, but I have a block from one of the cottage walls that is stamped '1863', the year the Christchurch Anglican Cathedral was built, which makes it one of the oldest buildings in New Zealand. It is made from limestone blocks quarried from the property itself, from one of the ridges above it, and the construction method is known as 'squared rubble brought to course' — the use of haphazardly sized blocks laid in courses, level, horizontal rows. Inside, it is rendered with limewash and cob — a mixture of tussock, clay, cow dung and water.

The same limestone blocks were used to build a plunge dip close to the woolshed behind the house. Before he died, Jim had got Bob McIlraith over to re-do the render lining, which was looking a bit sad. Bob's been a musterer all his life, but he's picked up a skill with stones and mortar from his father. He fixed up the dip, too. One day in early 1998 when we were shearing in the main shed, I had Bob look at the three old farm buildings. He agreed to come back and replace the missing blocks. Then he looked at the old house in the grip of a mass of ivy, the thatch on the roof gone and the whole lot full of rotten vegetation and bird droppings.

'You know, Jim always wanted to do this old thing up,' he said wistfully.

I knew this, because Jim had often said so.

'What d'you reckon would be involved?' I asked Bob.

'Dunno,' he said, scratching his head. 'Blimmin lot of work.'

'We've never been scared of hard work,' I said to him.

'No, Beverley,' he grinned. 'We haven't.'

I'd had the same dream of restoring the cottage myself, ever since I'd seen an old stone house opposite the showgrounds when I was judging down at the Lake Hayes A&P Show. It had been done up and its owners now used it as a tourist attraction. It seemed to me that Stoneycroft on Black Hills could have a similar future, if only we could fix it up.

Bob came back with his tools and did the shed walls. We

Top The cottage as it was before restoration work started.

Bottom Restoration of woolshed and granary started in 1998. Russell (left) and Warren (right).

kept talking about doing the cottage, but it looked a bit hard, but we decided to have a crack at the woolshed. Bob did most of the stone work, ably assisted by Brian O'Sullivan and Russell, and I did what I could. Warren Mason, another school friend of Jim's, did all the woodwork, using timber milled right there on the farm. Warren's a real craftsman. He was a farmer, but he had always wanted to be a builder. He used to watch the builders who were working on my new homestead up on the hill here, and tell me that they were 'real' builders, while he was just a woodworker. I assured him that what he had done with the old buildings — imagining what had fitted where the wood had rotted away, and restoring it so you'd swear it had always been there — was a far greater skill than just fitting a kitset together.

Once we had finished the shed to everyone's satisfaction, we tackled the stables. It was only once we were finished that — in August, when the ivy was bare and without leaves — we turned to the cottage itself.

At first it looked hopeless. The ivy had worked its way into the double-blocked walls, cracking them and making it impossible to pull the vines away without breaking off bits of the wall as well. You could hack and saw at it and get nowhere. Desperate measures were called for, so we roped in a mate of Warren's who had a great big logging truck. We numbered all the blocks, wrapped a chain around the main trunk of the ivy and lifted it off with the Hiab. Great slabs of the walls came away, too. When the dust had cleared, it looked like a giant, three-dimensional jigsaw puzzle, but the basic structure was still standing.

'Lucky we're not afraid of hard work, Bob,' I said.

Bob smiled grimly.

Off and on, over the next couple of years, Bob, Brian, Warren and Russell painstakingly reconstructed the walls. None of them, not even Bob, had ever worked with the rubble method of construction, but he learned on the job. They used only handtools — an escutcheon hammer, mallet, cold chisel and bolster, a trowel and a pointing tool — and stone from the site, and from around the property, where there are outcroppings of similar limestone. I used to wriggle blocks onto the transport tray of the Massey Ferguson tractor to bring them down — the sight of me being rattled up and down on the machine's hard springs earned it the nickname 'the bosom bouncer' amongst us all.

In an old photo of the Forrester cottage, in Jeremy Salmond's book *Old New Zealand Houses: 1800–1940*, you can see a round stone leaning up against the cottage wall. Russell found that very stone amongst the grass on the flats when he was sowing lucerne, albeit in two bits. When the walls were finished, I took enormous pride in leaning it up in the same spot as it is in the photo in the book.

The roof would originally have been thatched with the snow tussock that used to grow in the Hurunui, but that's in pretty short supply these days. So we decided to use ti-tree, and we laid it over corrugated iron so that we had the look, but we also had a properly watertight roof to protect everyone's hard work.

I helped muck out the interior. Quite early on in the piece, while rummaging through the artefacts amongst the bird manure and rotten straw on the floor of the house, I found a tiny leather boot. Whoever had made it clearly knew what they were about. Then I recalled that Frank Perrott had been a bootmaker by trade. He almost certainly had made this little thing. Funny how a man who was, by all accounts, capable of being so harsh could also be capable of making something so beautiful.

DOING UP THE BUILDINGS gave me a real stake in them, and in the land around them. I found myself devoting more and more of my time and energy to the farm. I think my vision was that the buildings could serve as a sort of living museum of the early settlers' way of life, and to this end I gathered the bits and pieces of antique farm machinery, carts and drays and assorted horsey things from around the property, and put them on display in the cottage and woolshed. I included a display of my coloured wool and yarn.

People began to visit, whether just stopping en route to somewhere else or (soon enough) by the bus-load, keen to look at the historic stuff and to get a sense of what it's like on a working farm. They were mostly tourists at first, but I've noticed more and more New Zealanders coming to enjoy the experience, too. Unlike days gone by — perhaps even 50 years ago — most Kiwis don't have much to do with the country anymore. So now I found myself adding 'operator of tourist attraction' to my portfolio of responsibilities! In March 2000, the Hurunui District Council flattered me with a Heritage Award for our team effort in restoring the buildings; they honoured me again in 2003, with a Heritage and Environment Award. I applied for and got Qualmark accreditation for the Black Hills Farm experience in 2005, the same year that the first of many weddings was held in the picturesque yard in front of Stoneycroft.

THERE WAS NO POINT in me rattling around on my lonesome in the big old house at Heatherdale. I would need to free up a house for a manager to help me with the farming. Besides, I was living in my in-laws' house. While it had been OK when it was

Jim living there and I was his wife, now that it was just me it didn't sit quite right.

But I didn't want to go back to the house on Karaka. It was too close to the road, and I wasn't sure I would ever feel safe by myself there. So I decided I would build another house on Perrotts, and the site I chose was across the creek and up on the hill overlooking Stoneycroft. From there, I could keep an eye on the old buildings, and glimpse visitors before they came up the driveway to the house.

What was quite nice about starting from scratch was that I could set up a house the way I wanted it. Whereas most people who get to call the shots when they build go for the aesthetic touches — what surface the kitchen bench should be, where the picture window should go — I was far more interested in getting the practical aspects right. With all my work with people with disabilities in grossly unsuitable houses, I wanted to make sure that my house would be comfortable for an elderly or disabled person to live in. Who knows, that person might just turn out to be me! So I ensured there was level access throughout, decent insulation (sheep's wool, of course!), wide doorways and no tight corners that a wheelchair couldn't negotiate. Not that I neglected the aesthetic elements: I wanted it to look more or less in keeping with Stoneycroft, and the 'look' I settled on was a rough mortared parquet appearance using special, sandstone-coloured fired-earth bricks from Swanhill in Perth. It was ready to move into on 10 May 1999.

With my block — Perrotts — taken out of it, Heatherdale was uneconomic. It had certainly put four Forrester children through boarding school, back in the day, with meat and wool prices right up there and with agricultural subsidies keeping costs down. But not now. It was harder and harder to make any money out of Corriedale wool. I took the hard decision to close the stud and disperse the flock. The Herefords were already gone: after Jim died, Doc selected the best dozen of the cows to run alongside his own stud animals.

The Heatherdale property and the homestead didn't pass out of the family: Ben Cassidy, eldest boy of Jim's youngest sister Rosemary, is there. The plan had always been that he would come back to assist Jim with the running of the farm once he had finished his courses at Lincoln and after he'd done a stint working in the United States. Jim's death brought all that forward. He came back and worked with Don and Russell and Bob. When Frank's estate put it on the market, Ben took it on in June 2000. His children are the sixth generation of the Forrester family on the farm.

Around the same time, I decided to lease the Black Hills. The lessees were Bruce and Jane Black: Bruce is a cousin of Jim's, and a nephew of Jan and Doc Sidey's, too. We tried that for five years or so, but when the Blacks decided to move on I put the Black Hills property up for sale. The Cassidy family, who with Ben had bought Mount Alexander (just behind Black Hills as I look at it from my place, going east towards the sea), bought the tussock block, a big chunk of Black Hills, at the same time. The rest passed out of the family. Amazingly — one of those circularities — the latest purchaser of one of the blocks in 2012 is Beau McRae, a descendant of the McRaes who farmed Glens of Tekoa, Robert Adam Forrester's first pastoral employer way back in 1863.

These were relatively easy decisions to make, and they were just reflections of the financial realities, really. I made a much harder decision back around 2004. I had worked as an occupational therapist for 33 years — and as the North Canterbury Country District Occupational Therapist for 19 of them — and pretty much loved every day of it. But opportunities were opening up for me in the coloured wool direction. I travelled to a World Hereford Congress in Argentina in 2004, and taking the time away from my practice proved to me that,

217

if I was to devote the necessary time and energy to farming, then I would have to give OT away.

Over the years, I had acquired various bits and pieces of gear with Lions Club sponsorship — a commode, a wheelchair or two, crutches, bed rails — and I hung onto these with a view to lending them out to people around the district who found themselves in need of them. I still do this, sharing them with the medical centres when need arises, and I still pick up bits of kit here and there when I spot them. Funds from the bus tours to the farm go towards buying new gear: people on the tours really seem to appreciate knowing that.

At last, I could devote my time and energy to the farm. I upgraded the fencing — well, when I say 'I', I mean 'we', but I helped a little; since 1999, we have renewed four kilometres of fenceline and put in a big laneway. I also did a lot of planting about the place, both practical things like shelter belts and ornamentals as well. I think we stuck in about 2,500 trees.

One significant change I made to the farm operations was to begin to use more environmentally friendly pest and weed control. I had always had a lot of sympathy with the biodynamic farming movement, and now that I was in charge I began looking into it. I liked the look of the Biogro certification process, which entailed getting soil and water tests done to ensure there aren't too many nasties latent in the land, and then using only certified animal remedies and pest control methods.

For example, we started drenching the sheep with a mixture of apple cider vinegar, ti-tree oil and garlic, and we followed up the devastation wreaked by RCD on the rabbit population with good, old-fashioned shooting and trapping. The remedies we use for other pasture problems, such as thistles and barleygrass, are the same ones that people would have applied in R A Forrester's day: elbow grease and a grubber.

Top Graeme Twose
shearing a black sheep.

Bottom Beverley classing
wool at shearing time.

9
WHERE THERE'S A WOOL

LOOK AT MY MOTLEY COLLECTION out there. I've got the ewes up close to the house so I can keep an eye on them. Most of them have got lambs at their heels now. The first round of lambing finished about three weeks ago, and the second cycle is just starting. I woke up this morning to find that one of my ewes had had two lambs overnight, but only one was getting anything much out of her. She's pretty young, and this is her first time lambing — so with twins, she didn't know which way she was pointing. But she had at least got them out. I've discovered this morning that her udder's only running on one cylinder, so I've had to take the smaller of the two lambs off her so that I can bottle-feed her. Jude's boy Lyall will enjoy helping with that job!

My flock are mostly Romneys, Corriedales and New Zealand Half-breds, with the odd Merino and Polwarth in there, too. But people are puzzled to see a flock of them, because they're the wrong colour. They're brown, black and at least 50 shades of grey. Some of them are several different colours: they're spotted, striped, mottled and splashed with various shades of natural colours. One or two of them even have pale blue-grey patches. Sheep are supposed to be white, aren't they? After all, that's where the saying 'the black sheep of the family' comes from: there's something suspect about coloured sheep.

In fact, sheep were all natural coloured before human beings started tinkering with their genetics, and before white sheep were even identified. Feral sheep are natural coloured.

The immediate ancestor of the domestic sheep is thought to have been something like the Mouflon, a goat-like animal that was indigenous to parts of Europe and Asia. They were mostly brown, and had a double coat — coarse hair for protection, underlain by finer, more insulating woolly fibres. There are signs that human beings first began domesticating Mouflons in the Middle East, in what are now Turkey and Iran, nearly 10,000 years ago.

By the time agriculture had spread as far as Britain around 5,000 years ago, it's likely that sheep much more closely resembled the animals we know, probably being quite similar to the Soay (the name means 'Island of Sheep' in Old Norse) sheep that can still be found on the islands of St Kilda, off the west coast of Scotland. The Soay is still quite goat-like, small compared with modern sheep, extravagantly horned and very agile. Their colouration resembles that of the Mouflon: they're usually brown or tan, with white patches under their chin, on their belly and rump, but they have no obvious outer coat and instead a fine fleece, which suggests that they are the product of selective breeding. The Soay is one of those breeds that actually sheds its fleece, and you can pluck them like chooks in the summer.

There is evidence that people were spinning wool in the Orkneys around 5,000 years ago, but the first fabrics woven from wool that are known to archaeologists come from Denmark, and date to roughly 1700BC. During the same period, the Phoenicians were apparently manufacturing dye from species of shellfish found only in the Mediterranean. There must have been quite a demand, because they managed to fish some of these creatures to extinction. And if there was a demand, it can only be because people were dyeing textiles, which means they were growing white wool. So we can safely presume that people breeding sheep were deliberately selecting for the white wool that certain genetic freaks in their flocks produced.

All the same, most sheep were coloured. One of the more ancient of the breeds with which we are familiar is the Merino, which seems to have arisen as a cross-breed between African and Spanish sheep, with some input from the Italian Tarentine breed. These may have been improved by inter-breeding with fine-woolled sheep brought from Africa by a tribe of the Berber people known as the Beni-Merines, from which the Merino breed's name derives. This last major genetic modification seems to have taken place around the twelfth century at a time when Spain was under the control of the Moors. There was a major Spanish textile industry based on the Merino, and, even after the Moors were cast out of Spain by the Crusaders, Merino flocks were jealously guarded by the Spanish nobility, such was the value of their fine wool and of the lovely, soft fabric that could be woven from it. Most early Merinos were light brown in colour.

It was the Romans who brought the ancestors of most British sheep to Great Britain, around 2,000 years ago. The Romans had a marked preference for white-woolled breeds. Subsequently, the English became expert livestock improvers, and the importance of their textile industry to the country's economy is indicated by the fact that the Chancellor of the Exchequer has sat on a seat known as 'the woolsack' since the fourteenth century. The breeds introduced by the

Romans and improved by English farmers all but displaced the ancient British breeds; the Highland Clearances of the nineteenth century were the last gasp of this process, with white, polled Blackface and Cheviots chasing the horned, coloured sheep traditionally farmed by crofters to outposts on the offshore isles.

New Zealand was settled by people in whom the preference for white wool was firmly established. And of course, if you were that fixated on the production of white wool, the occasional throw-back amongst the lambs — a splash of natural colour — was an aberration: the black sheep in the family. It happens quite a bit, because the coloured genes remain there. It's just they're recessive rather than dominant, as they were in ancient sheep. It's good for the likes of me, interested as I am in bringing those recessive genes to the fore and breeding coloured sheep. People are only too happy to pass you a black or coloured lamb, and they'll often drop them off under the cover of darkness. Funny thing is, you'll often see the same people proudly wearing a handknitted, hand-spun, natural-coloured wool jersey at the show, and they'll often ask after their own animal and be as pleased as punch if you've got it there on show.

I refresh the genetics of my flock from time to time. I'll put my black rams over a couple of white ewes from Sidey's, or I'll put a white ram over my black ewes to try to get a bit of size and to get the quality of the wool right. The resulting progeny are 'heterozygote' — carrying genes for both white and coloured wool — and you have to wait for two more generations to get coloured progeny.

I'VE ALWAYS BEEN A SPINNER, knitter and weaver, so I find it easy to understand the current revival in woolcraft. I find knitting incredibly therapeutic. I like calling it the new yoga. It so happens that just as I was getting interested in coloured sheep genetics in the 1970s, there was a bit of a resurgence of interest in doing things naturally: the farmers' market movement, and the insistence on knowing the history of food from paddock to plate were where one part of this movement was just taking off, as a reaction to the fact that there were children who genuinely believed that milk, for example, came from a cardboard carton or a glass bottle. The revival of woolcraft was another. And part of the attempt to return to more 'natural' ways of doing things was turning back the clock on some of the aspects of wool-growing that have become industrialised.

The first international congress on coloured sheep was held in 1976, and since then there has been a steadily growing movement for the rediscovery not only of traditional handcrafts, such as spinning, knitting, weaving and crocheting, but also for the raw material: coloured wool. By the time I had moved to the Hurunui, I was producing, and selling, garments hand-spun and knitted from my own naturally grown, naturally coloured wool, and I was shifting as many as I could make.

Things began to take off after Jim died. It sort of stands to reason: the main limitation on the scale of my wool-growing was the amount of time and energy I could put into it. It's another one of those things where the more you put in, the more you get out. There were other spinners and knitters out there, too, and so an increasing demand for my wool. Supplying others made more sense than making my own garments, because by the time you've carded and spun the wool — it takes 40 hours to spin enough for a garment — and knitted your jersey or whatever the project is (another

40 hours), you've spent 80 hours on it. The most you can hope to sell a nice, handspun jersey for is $250, so the hourly rate looks a bit on the measly side. Anyway, soon things were chugging along nicely to the point where I was given a Her Business Network Award for my wool-growing and marketing activities at the New Zealand Businesswoman of the Year function in Auckland in June 2003. But if I thought I'd hit the big time then, I had no idea.

THE VARIOUS BREEDS OF SHEEP produce different kinds of wool, from what's known as 'superfine' wool — Merinos are the most famous producers of this — to so-called 'strong' wools at the other end of the scale. Wool fibre is measured in 'microns' (or micrometres). The average human head hair is 100 microns. Merino wool tends to be between 10 and 20 microns, with most of it being around the 18 micron mark. Coloured Corriedales like mine tend to produce fibres from 22 to 30 microns; Polwarths — which are an Aussie breed developed in the nineteenth century from Merinos and Lincolns (friends of mine actually have Flock #1 in Australia) — are 23 to 25 microns; and Romneys are 31 to 39 microns. Obviously, the finer the wool, the finer the fabric that can be made from it. The New Zealand version of the Romney (which is actually quite different to its English forebears) suited our old methods of wool production, where strong wool was favoured and destined for carpet-making: machine shearing followed by a chemical scour.

For my purposes, much more care needs to be taken. To start with, I don't want to use chemicals in my wool, either in raising the sheep or in treating the fleece once it's removed from the sheep's back. This makes it hard to achieve the kind of shrink-resistance that industrial yarns tend to have: under the microscope wool has lots of little barbs on it, which is what causes it to shrink in the wash, with the fibres becoming snarled with one another and the strands therefore shortening. The industrial shrink-resistance processes tend to remove these barbs with chemicals — in effect, turning it into a different fibre from wool. My wool doesn't go through all that, so the propensity of my yarn to shrink is just one of the limitations that people need to be mindful of.

There's an upside to chemical-free wool, though. You'll often hear people say they can't wear wool, because it irritates their skin. Just the other day I had a discussion with a young man, well educated so far as I could tell, who insisted he couldn't wear wool. I pointed out he was wearing an Icebreaker shirt next to his skin.

'Oh, that's not wool — that's Merino,' he replied.

Imagine his surprise when I told him that Merino came from sheep! Merino growers call the finer fibres in their fleece the 'comfort' factor, and the coarser wool the 'prickle' factor. It's true that the coarser the wool, the more it is likely to prickle the skin. But quite a lot of the irritants that industrially produced wools have are down to the chemicals used in the scouring and dyeing processes. My aim is to grow wool on free-range sheep and chemical-free, and, as my customers will tell you, they can feel the difference.

IN THE LATE 1990s, Mum and Dad travelled to the United Kingdom and went to Hereford, which is where Dad's father, Sam Price, hailed from. They were half-hoping to track down

Price relatives, but didn't have much luck. Someone suggested they put a notice in the local paper, which is what they did. And about a month after they had returned to New Zealand, they got a reply from someone whose mother was a Mansel Price descendant. Mum and Dad went to the UK again and visited, and got to see the house in which Sam had been born. Sam's family apparently had no idea what had become of him, so they were able to fill in a big blank in their family history. Two of Dad's cousins came out for Dad's seventieth birthday, and took Noelene and me under their wing. They went to the Hawarden Ewe Fair, which they enjoyed, although the North Canterbury sun wasn't too kind on their pale Herefordshire hides!

Among the relatives Noelene and I later met in the UK were Brian Cole and his wife, Penny, cousins who lived in Watlington, Oxfordshire. They had an interest in wool: Brian had been an accountant out on the Shetland Islands for six or so years, during the time of the demise of the Shetland wool industry. He was there at the time of farmers reorganising themselves into a more artisan supply model. They are an imposing couple. Brian is very, very tall, with equally long arms, which might explain his interest in bespoke garments — you're not going to find much that fits his frame straight off the peg. Penny is very tall, too, and as slim as a pin. They came out to visit in 2004, and they were most interested in what I was doing at Black Hills. They reckoned there might be a market for my yarn, and maybe even for the garments themselves, over in the United Kingdom. I was excited.

They each took a hand-spun garment back to England which they also used as samples. I didn't think too much more about it, because there was plenty to keep me occupied. The World Congress on Coloured Sheep was held in Christchurch in November 2004, and I met lots of people from all over the world who shared my interests and philosophies with regards to wool and woolcrafts.

Our Price great-grandparents, Alice
Louisa Price (née Wright) born
2 October 1856, died 14 August
1908, and Mansel Price born 7 April
1844, died 25 December 1920. He
was a master carpenter and later in
life a publican.

That time around, they asked me if I would travel to Oregon in the USA the following May to be a fleece judge.

Janice Winter, a friend of mine from the Coloured Sheep Association with whom I've been to other World Congresses, ended up going with me, and it was widely believed amongst the Americans that we were a 'couple'. We were treated accordingly, and it took us a while to work out why! Ending up in the bridal suite at one of the motels we stayed at finally gave us an inkling . . .

The night before the competition, one of the stewards told me that I would be judging about 420 fleeces. I must have looked a bit stunned, because she told me I'd have two days to do it in.

'You'll have about 60 people sitting watching, and you'll need to talk to them all the time while you're doing it,' she said.

'Cripes,' I replied. 'I don't know if I can do that.'

Janice laughed quietly.

'You can do it. Of all people, Beverley, you qualify for that job.'

I drivelled on for two days, and that seemed to do the trick. Janice and I finished off by doing a skit for them, which they enjoyed. Must have done, because they've asked me back, and I'm supposed to be going to Maryland, too, this next year.

This, along with my association with Rural Women New Zealand, which has international networks, opened all sorts of doors. I've been to World Hereford Cattle Congresses since then, including in Scandinavia, Canada, Chile and Argentina, and I now belong to the Associated Country Women of the World (ACCW).

AROUND THE SAME TIME, I was working on an application to have Black Hills recognised as a 'New Zealand Century Farm and Station' — a scheme that was established to preserve rural heritage by identifying those farms that had been worked by the same family for 100 years or more. Applying entailed trawling through the historical records, reading all those old newspapers, diaries, deeds and letters and things, and I loved it. I was invited to the inaugural awards ceremony in Lawrence down in Central Otago in May 2006. The only trouble was, I was allowed to take a partner and I couldn't at first think who should come along. Then it dawned on me: for many of the past 100 years, Russell had worked on the farm — who better to go along than he? His wife, May, didn't turn a hair when I told her that I needed to borrow her husband for the weekend. After all, it was a running joke amongst the three of us that she was Princess Di to my Camilla, her husband dividing his time between their place in Waikari and my place up here at Black Hills. The three of us travelled and had a wonderful time. The Prime Minister at the time, Helen Clark, presented the award, a large brass plaque which I proudly display in the old cottage.

AROUND THAT TIME, Brian and Penny Cole rang and told me that there was a Stitches Trade Fair happening at Alexandra Palace in London in October. He reckoned he'd had a lot of interest in the jersey I had given him and the vest I'd knitted for Penny. It was another one of those moments where someone is urging you to do something outside your comfort zone, but I didn't have to think about it for too long. What did I have to lose?

I had no idea what would be involved in exporting a consignment of yarn. I drove down to the Christchurch offices of the Ministry of Agriculture and Fisheries (now MPI) at the airport to find out. The doors opened onto a wooden counter, beyond which were three men seated at their desks, listlessly toying with their keyboards. None of them acknowledged me, let alone got up. I couldn't tell who I was supposed to be talking to, so I addressed myself to all three of them, and told them what I was trying to find out. They looked at each other. One of them told me that his computer was broken, and he couldn't help. Another thought I should be talking to Customs and I needed to go into Christchurch city, but couldn't tell me where to go, as his computer was down. The third just plain ignored me. I drove back home, completing a 200-kilometre round trip, disgusted at how I had been treated.

Luckily, the next day my former sister-in-law and her husband dropped in for a cuppa on their way home from Hanmer Springs. When I told them about my experience with the bureaucracy, she said: 'Don't worry, Beverley. Our son's a freight-forwarder. He works at Camerons. He'll sort it out.'

And he did! It wasn't a simple matter, as it turned out that you could easily export wool in the grease, or finished garments, but almost no one had been exporting yarn still with its natural load of lanolin. It took the best part of a week to sort out. In the end, though, I exported my first-ever bale of 185 kilograms of coloured yarn — there were seven colours in all, all of them natural — on the 6.25pm Emirates flight from Christchurch to Heathrow on 6 September 2006.

It arrived in Watlington just over a week later: total cost just on $NZ10 per kilogram.

In October, off to England. Penny and I travelled by train down to London, visited the Marks and Spencer knitwear section, then went off to the England Knitting and Stitches Show, Alexandra Palace. It was a very big day. If you watch major British darts or figure skating competitions on TV, you'll know the Alexandra Palace, a wonderful grand building in the shadow of the British Telecom building (although ironically cellphones and EFTPOS machines only function intermittently there). We loved the whole experience — approximately 450 trade stands — exhausting, stimulating. There were handcrafts everywhere, with about 12,000 people attending a day for four days — that's a total of 48,000 people! Did people want wool? Did people want New Zealand wool? Did people want BlackHills natural coloured, undyed, chemical-free 100% wool? What about knitting yarn? Question after question.

Encouraged by the response to the product and what we began to call 'the brand', Brian and Penny did a tour of retail chain stores, testing the waters. The results were positive.

'I think we should start a business,' Brian said. 'A sort of a retail outlet and showroom. You can supply us, and Penny and I can manage it.'

I went hot, then cold.

'What would we need to do?' I asked.

'Well, you'd need to think of a brand name,' he said.

'Black Hills would be appropriate.'

Not that it *was* a simple matter, of course, negotiating all the unexpected costs and legalese and so on involved in finding buildings and then employing staff. Brian and Penny handled all of that, but they kept me in the loop every half-inch of the way. We all have two clocks, because of course while we're asleep in New Zealand they're just waking up in the UK. We do our talking in the evening, as my long day ends and theirs begins. Daylight saving makes for fun and games.

The whole of what you'd call the overseas 'business culture' is different to New Zealand's. We're more informal here, and while my rule has always been that you're never later than five minutes early for an interview or a meeting, in Britain you wait until a quarter of a minute before the scheduled time before you knock on the door. And where we have a solve-it-as-you-go creative problem-solving approach to things, in Britain there's a proper way — for everything.

We brought Simon Smith, a British filmmaker with 15 or 16 Hollywood films to his credit, out to New Zealand to shoot a promotional video — you can see it on my website — and that brought a whole suite of challenges along with it: getting music we wanted to use in the background written and copyrighted, and so on. Clare, a New Zealander in London, had previously worked on the Holmes Show with TVNZ. She did the voice-over for the DVD. We now had the New Zealand film with a Kiwi accent and from an English perspective! I later learned I had flatted with her mother's sister, who is still a mate from Country Girls days in the 1970s, and Clare's grandparents lived at Hurunui. Connections unbeknown! Incidentally, I just got a call the other day from Rural Women New Zealand, who had been asked by the organisers of an international agriculture day in Rome whether they had a video showing women involved in agriculture. So that film was sent off to them.

At this time, we applied for a Marketing Development Grant from Trade and Enterprise New Zealand. We were turned down twice, before I went to my MP and on to Parliament where we were successful in getting the legislation changed to include sole traders. However, I still did not get a grant.

The following October, the BlackHills showroom/shop opened in Watlington, Oxfordshire, within Brian and Penny's accountancy premises. It was a spectacular day, the shelves were filled with a kaleidoscope of natural-coloured yarns, garments covered the walls and up the stairs, lots of guests, lots of talk, newspaper interviews by the *Henley Standard*. The resulting article was titled 'Kiwis pull the wool over the eyes of the English'. We had arrived.

Two days later we had repacked and relocated to Alexandra Palace again and now stood behind the desk at the BlackHills stand at the England Knitting and Stitches Show. This time

I also sent over all the gear needed to set up our stand — it was cheaper to send the mannequins and displays and so on from here than it was to buy it all over there. Most of it was put together by Brian O'Sullivan, who helped with the work on the restoration of my buildings. It was ingenious stuff, too: all collapsible and fold-away for easy packing and freighting.

People like meeting the 'lady farmer', which seems to be part of the whole package, and for my part it was a marvellous way of getting to meet our customers and to hear at first hand what they respond to. Of course, there were lots of expat New Zealanders there, for whom the sight of Maori place names — Waikari, Hurunui — and the warm scent of lanolin sent them off on a sentimental journey.

It was, needless to say, marvellous exposure, and within a few weeks we had the first retail stockist of BlackHills yarn sign themselves up: a little wool and knitting shop in Stroud, Gloucestershire.

The business in England began to evolve as a result of the 100% commitment from Brian and Penny. A website was established, pattern designers and knitters contracted, knitting groups formed, and so much more. Most important was the direct service and the rapport with our ever-increasing numbers of customers, and Penny's skill in managing the shops and staff.

Over the past few years, I've shuttled back and forth between New Zealand and the UK. I've met lots of people from all walks of life, and feel as though I have become part of dozens of families around the world.

In February 2007, trademark status for 'BlackHills' was gained with the byline 'BlackHills — New Zealand Wool — Naturally'. Thought was also given to a label for marketing my garments, and 'Beverley Riverina — Fleece to Fashion' was trademarked.

I was now an exporter — and a fashion designer, too!

10
BY ACCIDENT AND BY DESIGN

I'VE JUST HAD A CALL from Selwyn and Michele Boler, who want to bring their endurance horses down this way, as they do from time to time, and keep them here for a while. Selwyn's the son of Lynn Boler, who was Dad's fellow hazer at the rodeo way back then. Lynn, Dad, Selwyn and I were in the same polocrosse team — I still wear my Tauhoa Polocrosse Club helmet when I ride the quad bike. Funny thing was, Mum phoned from Warkworth one time and told me that she had a hairdresser from up Kaipara Flats way who she liked to use, and blow me down if it wasn't Selwyn's wife! That's how we reconnected after all these years.

You seem to strike those connections everywhere. Mum and Dad used to farm next door to the Davie-Martins. Alan and Sharon were younger — our age, actually: Alan was at school with Noelene and me. We were quite close, really. Alan used to help Dad with cattle, and Dad helped him, and Alan was a bit like the brother that Noelene and I never had. Well, they shifted down here to Culverden to go dairy farming a few years ago: I can virtually see their place from mine. (You should have seen the locals when all this dairy farm gear began arriving on the hill on my farm! They thought I was going into dairy, and I had a bit of fun telling people that I 'like to keep my options open'. I was only ever storing the stuff for Alan.) Their three daughters progressively came down to join them. Rachael is a model, and has actually won a big beauty pageant or two in her time. She is a wonderful help in all aspects of my fashion parades, and even her children now help.

In 2011 or around then, I was due to host the Wool Equities people at Black Hills — lots of important men in suits from Wellington and wherever — and we put on a fashion show for them with my verandah serving as a catwalk. Jim Sidey's wife, Ramona, modelled: she's German by extraction, and long and lithe and looks the part, and their children also model my children's range. But Rachael added a whole new level of glamour. I asked Rachael if I could introduce her as a former Miss Auckland, and she muttered a bit about that and being Miss New Zealand. When I finally did my little spiel to introduce her, I described her as Miss New Zealand 1999. (Jim Sidey, not to be outdone, piped up and said: 'My wife is my Mrs Germany!') Well, these men, their jaws just dropped when Rachael walked out. She looked the part, all right. Funny that she was local, and yet connected to my past, too. I always believe that adage: you never get a second chance to make a first impression.

It's a similar story with the lawyers used to write our contracts and to protect the intellectual property. The garments are obviously our designs, but it's less obvious that some of the yarns we have developed are our own creations, too. I've done a lot of work in developing new yarns. It used to be much easier when Lincoln College had a wool research unit, where you could try out an idea by spinning up a few kilos at a time. It's much harder now, because to spin yarn at the woollen mills, you have to do a minimum of 100 kilograms . . . which is fine if it works . . . I've had some of my yarns copied, and that annoyed me. I now use the services of several lawyers, including one in London, whose partner I knew from a baby. His parents, Harley and Peggy, farmed up the Old West Coast Road. One day I was taking Mum and Dad for a drive up that way and we saw Harley leaning over his road gate. We stopped to say hello, because I knew Dad and Mum also knew him, with them all being Angus breeders.

'Hi, Harley. How's Peggy?'

'Oh, she's up at the hospital.'

'Is she not feeling too good?'

'She just had a baby. A little boy.'

'What have you called him?'

'Angus.'

Angus! What else?

Angus Jenkins's partner, Pithvie, is an intellectual property lawyer for the Holdson board-game people, and has been ideal for my purposes. Both he and Angus have also had some amazing ideas for men's garments, which I've been able to develop. So, you see, the local connection stretches as far as London!

I try to keep things local as much as I can, and the amazing thing about this district — maybe it's true of New Zealand generally — is that the people here have the most incredible range of skills. Jolene Gardner does my computer work. Originally from Gore, she did a diploma in architectural design, then a course in digital graphic design. Then, after having a family, she has set up as a freelance designer. She does all my graphic work — labels, signage, the bands that go around the yarn, everything — maintains my website and Facebook page and handles the mail-order side of the business. If it was left to me, with no understanding of computers and that stuff, and an exponent of the Columbus 'search-and-land' method of typing, Lord knows where I'd be.

Another member of my local talent pool is Susanne McLaren, one of my knitters, a Swiss girl who speaks about seven different languages. I get her along when the farm tours come onto the place. And when I delivered a lecture in French while in Paris just a few months ago at the World Sheep Congress, I didn't have to go any further than Waikari to find Emma, who has a PhD in French translation!

Vicki Sharp is another prime example. Vicki is a brilliant knitter, but she also has the amazing ability to write patterns. We used to use Tracey, a Somerset software designer who had a flair for writing patterns, too; we had others along the way — Caroline and Chris — but we soon realised it was just as cost-effective for Vicki and I to develop the patterns here. We'll get an idea — whether it's my idea, or something Penny has come up with from a customer over in the United Kingdom — and we'll kick the idea to and fro over a week or two, during which time Vicki will fax it to me. We'll decide what yarn to use, and what design and so on, and then Vicki will somehow turn it into a pattern which Penny or Jolene will type up, then we'll sell it locally and in the shops in the United Kingdom or the United States.

This entails getting your head around the differences between metric and imperial needles and yarn sizes, and so on, which Vicki seems to be able to do. Not only that, but most knitting patterns are for smaller dress sizes, and we live in the real world — and Vicki has the ability to work out how to scale a design up to produce a pattern for larger sizes, too. I'm so lucky to have these people with their skills at my disposal.

Anyone who's in business knows the value of a good accountant and bank. I have the best accountant for the job keeping an eye on me and my concerns: Craig Hastie is David

Hastie's son. He helps me with the balance of expenditure versus income, and keeps track of GST and PAYE return dates and what-have-you — all that complex stuff that would simply bog me down otherwise.

➤——————➤

THE 2007 WORLD CORRIEDALE CONFERENCE was held in Christchurch and was notable for the unveiling of a statue commemorating the work of James Little and other pioneers of the breed. It was also where, on 27 May, I launched both the Beverley Riverina fashion label and the BlackHills yarn brand. The function was called 'A Night in Paris' and was held in the Limes Room of the Christchurch Town Hall. It was professionally produced and presented by fashion-world stars Pranay Baidya and Andhy Blake, and featured models strutting their stuff in woollen garments designed by local legend Barbara Lee, and . . . me! My garments were worn with GAS Jeans, which is co-owned in New Zealand by Dan Carter. It was sort of surreal, but a wonderful evening at the same time. We have now done lots of parades, including one we were invited to do at Style Christchurch during Show and Cup Week.

Rough sketches of garments are made and, together with the very talented Vicki Sharp and Susanne McLaren, are knitted. Brian O'Sullivan's wife, Julia, does all my sewing, making all the net underskirts, trimming the garments with their velvet ribbons and frilly bits of embellishment. Since I'd got to know Brian through his work on the buildings, I've become close to them both. They were running coloured sheep, and what a bonus it was when I learned how clever she was with a sewing machine! She has been my right-hand-man at all the fashion shows since — whether it's with all the glitz and glamour up in Auckland or something we've put together as a fundraiser for the Culverden Fire Brigade — and she has quite a knack for knowing which shoes go with what bags and that sort of thing. She now calls herself my personal assistant, and she surely is. Well, all the other designers had one, and she reckoned I needed one, too.

Diane Nurse from Hawarden — who has been with me 13 years helping out on the farm — is now also working with me on finishing garments and yarns, and she's been vital to the exporting, stocktaking and dispatch side of things ever since, as well as helping out with the bus tours. She is a godsend.

Suddenly, I was a globetrotter. I received an invitation to travel to China as part of a delegation of mid-micron farmers visiting woollen manufacturing plants and textile training institutions in that country, and we attended the Wool Awards of China, run by Qifa, one of the largest Chinese hand-knitting yarn manufacturers. Back to the farm, and then an opportunity came up for an extended visit to the United Kingdom, Scandinavia and Europe, visiting potential customers and collecting feedback on what we were doing well with the yarn and brand, and what I could be doing better, while ferreting out new designs and designers for the fashion side of things.

October 2008, I was back at Alexandra Palace for the England Knitting and Stitches Trade Show, which was becoming an annual event, this time armed with a nice, glossy 'look book' for the Beverley Riverina fashion collection. Penny and Brian moved on to exhibitions around British Wool Weekend in the pretty little town of Harrogate, and then another in the Birmingham Exhibition Centre. It was all great exposure, and they got our yarn in front of lots of people who were in a position to assist the business.

Back in New Zealand, I was ready to put my feet up. But I'd only been back 11 days, and there I was helping the Cattle

Committee with a working bee for the Canterbury Show. I was in a room with my mate Nancy Gardner, putting together all the trophies, ribbons and cards for cattle prizewinners, when Rae Findlay put her head around the door.

'Beverley, we want you to knit us a jersey,' she said.

'Oh, yes,' I said. 'What sort of jersey?'

'We'll leave that up to you. But we need it pretty quickly.'

'How quickly?'

'Well, we'll need it in about 12 days.'

'Crumbs! What's the big hurry?'

'You know Princess Anne is coming out for the Commonwealth Agriculture Conference? We would like to present her with a gift, a jersey.'

Nancy's husband, Peter, had been at school with Jim, and Jim had died on Peter's birthday. 'Good grief,' Nancy said, when Rae had gone and I was sitting there, half-stunned. 'What on earth would Jim have thought of that?' I was asking myself much the same question. And I was reflecting that, while it was a huge honour to be asked, it was a lot of work in a short period of time. And I'd have to work on my curtsey.

I rang Vicki Sharp, seeking help. Of course she would! Heather Lilley had already carded and spun up some BlackHills fleece into yarn, so we selected a mixture of Moorit Corriedale and Romney — a rich, coffee-brown shade. My idea was that we would do a version of the traditional high-country musterer's jersey, with the round neck and a drop shoulder, so that we would be incorporating some of the proud tradition of New Zealand sheep farming along with the very best of its produce. We were supplied with Princess Anne's basic measurements, and we got stuck in.

We had to go like mad, though! It took 42 hours to completely produce. Then I had the thrill of attending the afternoon tea held in Her Royal Highness's honour, where I was introduced to her and got to chat to her about many topics — the wool

243

Top Diane Nurse with BlackHills yarn.

Bottom Beverley and Nancy Gardner at the Canterbury show where Beverley was asked to produce a garment as a presentation gift for Princess Anne when she attended the Commonwealth Agriculture Conference.

industries in New Zealand and Great Britain, the Christchurch Show and my own sheep exhibits, my business, of course, and my involvement in the Campaign for Wool, of which her brother Prince Charles is Patron. Afterwards, there was a formal ceremony in the central arena, in which I got to present the jersey to her, all done up in red show ribbons. I think I got the curtsey right.

'Is this really for me?' she asked, as I gave it to her.

'Yes, madam. It's really for you,' I said. 'It should fit, because it's made to your measurements!' And I told her all about the wool we'd used, where it had come from, and the style of jersey chosen. We also gave her a document with a description of the process by which the garment had been made, and a World Coloured Sheep Congress Proceedings book to read on the flight home.

She kept stroking the jersey, as she looked me right in the eye and said: 'Thank you very much.'

I didn't see her do it, but apparently as she was leaving the stage, she buried her nose in it and inhaled the wonderful lanolin smell that only natural handknits have.

BY MID-2009, I was sort of living in two worlds. This was starkly illustrated on 15 May, when I finished up attending the opening of a new shop, BlackHills Yarns at Henley-on-Thames (famed for its annual rowing regatta). Then I chased the sun around the world on an airliner, and was in New Zealand in time to see it setting on yet another 15 May in Wellington, where I made my weary way to Parliament for the announcement of the New Zealand Enterprising Rural Woman of the Year. I didn't win, but being runner-up had its own rewards. The phone started ringing off the hook with invitations to go and speak to this group here and that conference there about our work. And in December of that year, this included a live interview on Canterbury Television, in a building that tragically isn't there since the February 2011 earthquake. That was all on top of visits to the British trade fairs and a tour of England and Scotland promoting BlackHills yarn. By now, our products were selling through six outlets in the United Kingdom, and there was interest from others.

I was conscious of my double life again on 10 November, when Beverley Riverina was one of 12 labels to be selected for the catwalk in the fashion parade 'Style Christchurch' that wraps up the Christchurch Show and Cup Week. I sat and watched the young men and women — New Zealand and international models, as it happens — presenting my garments, and reflected that only that morning I had been using the shovel to keep clean the pen in which the sheep I was exhibiting in the Royal Canterbury A&P Show were housed. There's nothing like shovelling manure to keep it real!

But for me there's no doubt that the most important event of my whirlwind 2009 calendar was 12 September, when a whole bunch of people from the district gathered to celebrate Russell Hassall's eightieth birthday (and 60 years' continuous service to the Forrester family and their farming enterprises). Russell had always been an important part of operations, but he was — and still is — vital to mine. He must have wondered what he'd struck when Jim died and I had to take over. He was an oracle on how the farm was run and the fount of all knowledge, but despite all that he was receptive to doing things differently. Not all at once, mind you.

I remember once suggesting that we try a short-rotation crop I'd been reading about in one of our paddocks, before we put it in permanent pasture.

'Oh, no, no, no,' said Russell. 'That's not how we do it. It's been tried and it never worked for us.'

I left it for a week or two, and then brought the subject up again.

'You know that crop we were talking about, Russell? I know where I can get hold of some seed.'

'I don't know,' he'd say. 'We tried it once.'

'When was that?'

'Oh, 10, 15 . . .'

I left it for another week or two. Then I presented him with a sack of seed. He looked sceptical, and not too happy, but he drilled it and did his usual beautiful job of it.

A few weeks later we were in the Star and Garter Hotel and someone said, 'That crop you've got growing in the road paddock looks good.'

'Oh,' I replied, 'Russell put that in. Looks great, doesn't it?'

And Russell beamed and was fully on board.

No matter how far I have travelled, no matter how glamorous the events I have attended have been, I have always tried to keep my feet on the ground and remember that it has all been a team effort. Without Brian and Penny in the United Kingdom, there would be no export business and no UK wholesale or retail sales. Without Julie in the United States doing the same job of negotiating the quirks and differences of American business culture for me, there would be no sales in the United States and Canada. Without Diane tirelessly loading crates with yarn and ticking the reels off on her clipboard, Penny and Julie would have nothing to sell. Diane's a trojan. She labels yarn, keeps an eye on quality, does the stocktaking and sources fleece from the shed to supply replacement colours, and she keeps me on track emotionally as well, as required. She's always available to take my hand and give me an opinion or an option.

Without my team of designers — Vicki Sharp, who knits and does the commercial patterns and catwalk garments, and

Top Julie Miller, our
agent and distributor
in the US and Canada.
Bottom Vicki Sharp
with one of our knitted
garments.

Susanne McLaren, who, besides speaking all those languages, is also really good at designing and knitting the flamboyant catwalk, artisan and one-off, limited-edition garments we do under the Beverley Riverina label — and my sister Noelene does some, even my 85-year-old Aunty Jean in Auckland has done a garment for me — there probably wouldn't be any garments. Without the crew at Bruce Woollen Mills and Greenacres Fibre Processing, there wouldn't be any yarn on the spools. Heather, who lives in Balclutha these days, does hand-spinning when I want to be able to say a customised garment is hand-spun and handknitted.

I often say: no sheep, no wool; no wool, no business. Without Russell and his expertise, acquired over a lifetime of working on the same land he's working now, there would be no fleece to turn into yarn in the first place. Because, of course, it's not just a matter of turning sheep out onto the pasture and letting them eat grass and grow wool. Sheep have to be happy and healthy to produce a good fleece. You can always tell when a sheep has been stressed, because the wool has breaks — brittle sections — in it, which can lead to 'cotting' (the formation of clumps of wool in the fleece) or tenderness of the fibre that makes the work of the carder and spinner impossible.

And the woolly blighters are also subject to a range of other stresses: feed quantity and quality (which are affected by rainfall or the lack of it, and soil fertility), parasite burden, and pregnancy, to name but a few. And how the sheep are handled in the period immediately prior to shearing is critical, too: the fleece is prone to become contaminated with vegetable matter — twigs, hay, seeds and grass-stains — and by faecal stains, all of which make it pretty much useless for textile manufacture. That's why you want help on the job when you mean to produce high-quality wool. I'm so fortunate to have my quiet, calm Russell, and conscientious Diane, too.

ONE OF THE MANY, many people I have met in my recent adventures in wool is Michael Mellon, who worked as Marketing Manager for Coca-Cola in Hong Kong. Michael lectured in international trade marketing at Lincoln College and then the University of Canterbury, before retiring to pursue his own projects. He threw himself into a number of ventures, all of them motivated by his belief that, if New Zealand is to prosper, it has to have a vision beyond its traditional role of grower and exporter of primary produce.

In an age where many of New Zealand's old rural enterprises were being bought up by the corporates who have a sure grasp of the price of everything but no idea of its value, Michael threw his weight behind the value-add side that some of these old businesses brought to New Zealand agriculture. He was part of Geraldine's Barker's Fruit Wines, and Bush Road of Mosgiel, a supplier of fresh salads to supermarkets, and believed that similar marketing could be done with natural-colour chemical-free mid-micron wool. Michael is dedicated to trying to make it happen, and has become a fan of what I am doing. I'm very lucky to have Michael as a business mentor and be part of the Wool Advancement Group.

It was through Michael that I became part of a group of wool-growers that rescued Quality Yarns in Mosgiel in 2012, renaming it the Bruce Woollen Mills. The mill had been founded in 1897, back when the district in which the town of Milton stands was called Bruce County. Way back then, a group of sheep farmers had decided it would add enormous value to their operations if they formed a cooperative and built a woollen mill to turn their fleeces into yarn. So that's what they did, and the Bruce Woollen Mills and the imposing brick building with its traditional saw-tooth roofline were born. It was a big employer for much of the twentieth century, drawing workers from as far afield as Kaitangata, Balclutha and Lawrence.

In 1963, it was taken over by Alliance Textiles, and then by Quality Yarns in 1999. Under Quality Yarns' management, the size of the workforce was scaled back to just 19, from nearly four times that in the mill's heyday. And then, in what was turning out to be the twilight of New Zealand manufacturing in general, the business began to be run down and went into receivership in 2011.

What the Woollen Mills represented was a whole lot of specialised machinery, and a skilled workforce available who could operate it. The rationale for adding value to New Zealand wool products is as sound today as it was 100 years ago. Wool Equities Ltd — the corporate reincarnation of the old New Zealand Wool Board — took a 77 per cent stake in the business, and myself and 10 others became investors as well. That's where BlackHills yarn is spun these days, although I also use Greenacres Fibre Processing Burnham for smaller quantities of semi-worsted yarn. Before Diane labels and packs the yarn we receive from the mills for use in New Zealand or for export, she has to weigh it to make sure it's right. There's no use sending a 200-gram hank of yarn to a customer who discovers it weighs only 190 grams. Some of these then pass to Rowena for the new design — an adult jersey pack.

REALITY HAS A WAY OF intruding even on double lives. I had attended the North Canterbury Business Awards function on 3 September 2010, and received an award for niche marketing. We were in pretty high spirits, and the 10 of us had a wonderful night. Later, at the home of Stewart Johns (Jim's old schoolmate), at a bit after four in the morning, there was an almighty crack and crash and a horrible, sickening rolling. The lights didn't work, and we stood blinking in the beam of a torch that Stewart managed to find as the phone rang with people calling to ask if we were all okay.

Like most of Canterbury, we were lucky in that September quake, although a friend of mine in the Coloured Sheep Association lost her beautiful, two-storey home in Kaiapoi. We'd had a meeting there only 10 days before. A couple of days after the quake, I went to the Rescue Centre in Kaiapoi, representing Rural Women New Zealand, as locals were running short of basics like nappies, tea, sugar and so on. I did a bit of shopping and went down to their building in Kaiapoi, where there were water tankers parked so that people could fill plastic containers. It was like nothing I'd ever seen before. I delivered my items, and worked many hours with a clipboard, signing people in and out of the building.

Of course, it was nothing compared with what was to come. In February 2011, I had gone into Christchurch for the funeral of a young family friend we used to know in Warkworth who had been killed cycling on the Port Hills. He'd stayed with me for six weeks when he came down to study at Canterbury, and I represented our family at his funeral. Afterwards, I was in the central city with the idea of getting my grandmother's engagement ring re-sized at a jeweller's. Something made me go into the Cathedral, but I wasn't in there for long: I got a panicky feeling, and had to get out. The following Tuesday I was back down with Joy and Fred Morris with our standard-bred horses for sale, and I stopped in at Briscoes on Papanui Road. Just after 12.50 in the afternoon, all hell broke loose. We dropped to floor level. Stock was falling, the building was groaning and flexing. When it stopped, we all got out and the aftershock hit: we just sat in the gutter and watched the lampposts whipping this way and that like poplars in a high wind. It was unreal.

I carried on to the showgrounds, because our horses were due to go under the hammer. It was all confusion there. The sale soon stopped when word came over the public address that the Pyne Gould Corporation building in the centre of town was down, and that there had been serious injury. It was time to go home. Once the horses were safe, all I wanted to do was get home.

It took me an hour to get to Yaldhurst to join up with Joy and Fred again. They advised me not to try the Waimakariri River Bridge on State Highway 1, on my way home, so I headed up the back way through the gorge and back through Oxford and Rangiora. When I came to the little Waimakariri gorge bridge — it's only 50 metres or so — it took every bit of my courage to cross. I'm sure no pioneer ever faced one of these rivers lying between them and home with more dread in their hearts. It's the longest 50 metres I've ever driven. I dropped in on Stewart at Amberley, as I was exhausted. By coincidence he'd been on his tractor and, just before the quake struck, my black sheep had all gathered in a huddle in the middle of his paddock with their heads down: they obviously knew it was coming. I had three cups of tea with sugar — I don't usually take sugar — and then pressed on northward to my farm.

Apart from the power and phones being out, everything was okay at home, but the drama for us wasn't over. About half past 10 that evening, there was a knock on the door. I was nervous about opening it, because the power was out and I had no lights. But it was the young Maori chappie who had done some tree planting for me.

'There's a lady not very well at your front gate. She's had a heart attack. We can't get cellphone coverage, so can you phone the ambulance service?'

We summoned an ambulance, and it duly arrived and took

the poor woman away. We heard she lived, but, like lots of people you help out on the road, I often wonder what's become of her since. The effects of the quake linger. A couple of months after the quake, I was in Tasmania for a conference, and I had to force myself to get into the lift of the hotel — and I couldn't get out of it quick enough when it reached my floor.

There were other upheavals in my life around this time, too. On 26 January 2009, my dear dad Mansel died. He was a support and an inspiration to many, but none more so than to me: the farmer and rodeo rider who had been advised to keep clear of horses and rough ground and hard work 50 years before he suddenly died. Amazingly, he was survived by his own mother-in-law. Olive, who had turned 101 the previous November, was in amazing health for her age, and had been widowed for 17 years. It was one month later that Nana Olive died — Mum lost her husband and then her mother in a month. Along with Mum, Olive taught me to knit and assured me that it would prove to be a useful skill in later life. She lived to see just how right she was!

Three years later, my mum Ona died just short of 86. Noelene and I took turns caring for her in her final illness. Just before her funeral, we were contacted by the Warkworth RSA and told that they intended to extend full honours to her farewell — a rare tribute, and one usually reserved for returned servicemen and servicewomen. No less than my dad, Mum made me what I am. She was a force of nature, and quite happy to ignore advice on what was proper for a woman to do.

Where would I have been when Jim died if I hadn't grown up believing women could do farmwork? Ona was still buzzing about the property on the quad bike and grubbing thistles until ill health caught up with her only three months before she died. Besides the RSA, there were people from every part of the Warkworth community at the funeral to

254

farewell her. Mahurangi College was well represented: she had continued teaching shorthand and typing, and had adult literacy students after she 'retired'. You'd go a long way before you found anyone as worthy of being called a model of Kiwi womanhood. In everything I have done, I have tried to live up to her example.

LIFE, OF COURSE, GOES ON, and in my case it's only got crazier. A few months after we laid Mum to rest, I was packing off some of my garments to be shown at the fashion parade at New Zealand Wool Week in Auckland, which was attended by the Patron of the Campaign for Wool, HRH Prince Charles himself. I wonder if he knew his sister already had a BlackHills jersey? I know she wears it, because every so often someone will tell me they saw her in it in a photo in an English magazine.

The big break in my fashion-designing career had come on 21 June 2012, when I attended a BNZ Partners function at Russley. It was a terrible night, winter solstice and dark. When I got out of the car and tried to hurry across the unlit car park, I tripped over a concrete ledge and went down in a heap. I broke my glasses, my clothes had a tear, and, like my face, had blood on them, and although I had a quick spruce-up, I had to sit through the whole glamorous event in that state. Afterwards, when we were standing around, I was talking about how much I admired Trelise Cooper. Rachael Palmer was there with me, and she used to know Trelise through modelling. When he saw the state I was in, the BNZ manager was mortified, and asked if I'd like to meet Trelise.

Well, before I could say anything, I was shaking Trelise

Cooper's hand. She commiserated with me about my tumble in the car park, and we got talking. She was really interested in me and my work, and she asked lots and lots of questions. Then said, just like that: 'Why don't you come up to New Zealand Fashion Week this year?' I thought she was just being nice. Well, she *was* being nice, but she meant it, too, and before I knew it I was booked to go. The next year Beverley Riverina Handknits was selected as part of the 'BMW New Generation' section at New Zealand's premier fashion event in Auckland on 6 September 2013.

The other two contestants and Rachael and I flew up to Auckland three weeks ahead of the big day for what they called a fashion 'boot camp', where we were treated to a series of workshops on marketing and branding by people who were already successful in fashion, plus the CEO for BMW New Zealand and some top-flight fashion people. It was only then that the three contestants were each shown the nice, shiny new silver-grey 3-series BMW for which we each had designed a 'car wrap' — a decal that looked like a garment to be applied to the car. The car looked great when it was done! The three cars were used to ferry people to and from Fashion Week events and generally to raise the event's profile around town. They certainly managed that.

Rachael was along on the 'boot camp' with me, and she stayed on to help me select the catwalk models we'd use. About 500 candidates had been flown in from all over the country, and paraded, one by one, under the critical gaze of 50-odd designers. Maybe I'm not a bad judge of sheep, and I have an eye for stock, and judging good conformation in sheep and cattle, but I wouldn't have had a clue what to look for in a model. Rachael did most of the choosing.

The big afternoon at Fashion Week was one of the scariest of my life. I had the butterflies something terrible before the models took to the catwalk, but, as soon as the first model appeared in the blazing rectangle of light at the start of the polished-concrete catwalk, the butterflies coalesced into a single lump in my throat, and I watched the show on the backstage TV monitor through tears of exhilaration. It was hard not to get caught up in the excitement and the glamour of it all.

The icing on the cake was that our garments — Beverley Riverina Handknits, made of wool grown naturally on Black Hills and crafted by hand by Susanne McLaren, Vicki, Noelene, Mum and me — got a very positive reception, both from the crowd and from the people writing about it. We even got onto the TV news, once at 6.30 and then again at 10.30. I had set out to show that woollens weren't all about chunky jerseys and sensible skirts. We had garments like a full-length figure-hugging evening gown, a roll-necked dress cut just above the knee and accompanied by a walking coat, and another backless crocheted number with a bold, black fern on the front. I also had menswear — I was the only one at 2013 Fashion Week to have a full collection of natural-coloured chemical-free hand-knits — maybe the first time in Australasia for such a collection.

As I sat there, I remember thinking: if Mum and Dad could see me now!

ENCOURAGED BY THE SUCCESS OF the Fashion Week foray, the Wool Advancement Group offered to assist me and another yarn producer, Marnie Kelly from Alexandra — who works with possum, Merino, angora and silk — to attend the Vogue Knitting Live Fair in New York in January 2014. This is a trade fair that shows yarn and knitting products to buyers from all over North America. New Zealand had always had a bit of a niche in that market, but, with the combination of the global financial crisis and the persistently strong New Zealand dollar, this had fallen away. It was hoped that Marnie and I could go some way to restoring the country's profile in wool.

It was exhausting, but what a privilege! In November, Diane and I, with Marnie, packed a shipping container with yarn, patterns and knitwear — including some of the garments I had shown at Fashion Week — and all the equipment for display. We followed it across in mid-January 2014, collected Julie Miller in California, and set up close to Times Square, New York, where the show was to be held. The show was less well attended than I had expected, with about 5,000 people through. But they were all fascinated by our products. Once again I met amazing people and made good business contacts.

The most interesting conversation I had was with a teacher from a school in the tough suburb of the Bronx, who came through as I was replenishing my shelves on the Sunday morning prior to the doors opening to the public. We got talking when she fingered a staple of wool and asked me whether we had to kill the sheep to get the 'fur' off the skin. You wouldn't believe how often I get asked that question in the United States or in Europe. As I worked, I organised her to watch the short BlackHills DVD of sheep happily grazing at home and getting shorn, and the wool being worked into yarn for knitting and then made into garments. She was enthralled, and in turn told me all about where she taught, and the lives of her students, at least one of whom would come along each week in shock because some friend or family member had fallen victim to violent crime. It sounded just dreadful. But she was hopeful that by taking samples of my wool and telling them about how it was produced, she could give them a glimpse of a gentler world, far away from the one they knew, where an armed guard stood at the entrance to the razor-wire perimeter fence enclosing a school. I hoped so, too, although I haven't heard from them. I have this awful feeling that something may have happened to that teacher.

One evening, we met with an old friend of Julie's who is now CEO of the philanthropic organisation Broadway Cares. We had a wonderful evening, and he now wears a BlackHills handknitted jersey in the New York winters. At the show, I also met up again with Nicky Epstein, whom I had previously encountered in the trade shows in Britain. Nicky is a global knitting phenomenon, who not only supplies yarn and knitting products, but has also produced many dozens of pattern books of garments for people, pets and even Barbie dolls. She's written about the history of wool and knitting, and has led tours of world sites that have significance in the history of wool and textiles. She also teaches on cruise ships.

Nicky is a kindred spirit. She invited Julie and me to visit her and her husband, Howard, in their Manhattan apartment. It was like nothing I've ever seen — 25 storeys up, and still surrounded by the glass sides of other buildings. There was one gap, visible through the dining-room window. Howard told me that this used to be filled by the Twin Towers. Where we were discussing it was where he had been standing on the very morning of September 11, when the second plane struck, right before his eyes. Julie and I stood in awe. I felt ill.

BACK HOME, EVERYTHING WAS DRY, but it was lovely to rest my eyes on the hills after all that glass, concrete and steel, and the grubby plastic and rayon of airport departure lounges. My life might be busy — sometimes it seems too busy — but all it takes are a few quiet moments back here, in the heart of the Hurunui, with my animals around me, a cuppa beside me and the quiet click of my knitting needles soothing my soul, and I feel ready for it all again.

We have a bit of gap between town and country in New Zealand. It's probably getting larger, going by the response when we take our farm animals down to Christchurch for the Christmas children's nativity play at the Anglican Cathedral. My mate Helen is behind this scheme, and we've done it for about eight years now. I feel thankful that I'm not from a society that no longer knows what wool is and where it comes from, let alone how to protect its children from the horrors of the world without resorting to razor wire and guns.

Beverley (right) and Helen (left)
taking sheep and other animals to the
Christchurch Anglican Cathedral
for the annual nativity play and
children's Christmas service.

EPILOGUE

LOOK AT THAT. It's a year since I sat down to tell this story, and a whole new generation of lambs has been born. The little lamb that I put on the bottle is doing well, and weighs much more than her twin now. Lyall calls her 'Kahlua', which is quite a good name for a wee coffee-brown thing like her. He calls her sister 'Bailey' — I don't know where he got the names from! When Noelene was last down, she christened them 'Alice' and 'Madeleine'. Oh, well. They can wear two names each comfortably enough.

Normally, things would quieten down a bit now after lambing, but I've got plenty on my plate, as usual. Right now, we're working on a scheme to sell knitting packs to people visiting New Zealand on cruise ships. They'll contain patterns, needles, yarn and instructions — everything someone would need to knit a beanie, a scarf or a pair of socks. And I've been spending plenty of time out in the office at night, after the farmwork is done. I had to do something about my GST the other night, and rather than turn the office lights on (which can distract the truck drivers passing on the main road), I thought I could get to my desk in the dark — and I stepped on a blimmin mousetrap! Besides the routine paperwork, I've had my homework to do. While I'm a qualified occupational therapist, the only qualifications I have in business are from the School of Hard Knocks. Lately I decided I'd change that, and I'm part-way through a course offered by the City and Guilds Group of London.

I'm also trying to work out how I can get my sheep to the Amberley A&P Show. As luck would have it, I'm supposed to be going up to Auckland for the *Next Magazine* Businesswoman of the Year Awards, which Rural Women New Zealand have nominated me for. And I was persuaded to apply for a Nuffield Scholarship, a few of which are awarded each year to help the winners assume 'a leadership role in agricultural business and communities'. I think I'm probably a bit long in the tooth for that, as it seems tailored for younger people and those with university qualifications, but if I got to the interview stage it would require me to be in Wellington the night before I'm supposed to be at Amberley, and then somehow judging an hour and a half down the road at Ashburton at 9.30. I've been in the habit of taking my newly broken-in sheep to a quieter show like Amberley before I toss them into all the noise and bustle of the Canterbury Show — all the people, the talking, the music, the bands and so on — to get them used to it. I'm not sure how I'm going to make it all work.

Last year, I won a trophy that I didn't even know existed — for the most points around the local shows. I would quite like to defend my title.

I have been lucky with the support of organisations like Rural Women NZ, North Canterbury Enterprise and the Bank of New Zealand, which have put me through a whole lot of courses on branding and marketing, invited me to seminars, and plugged me into a whole lot of networks. They also put me through Canterbury Chamber of Commerce courses on technical matters such as bills of lading and the various bits of legislation governing exporting. And I was nominated for the Chamber of Commerce Air New Zealand Cargo Export Award. Connections, that's what it's all about — whether it's family, or friends, or business associates, or that peculiar Hurunui (and New Zealand) mixture of all those categories.

>——————▶

IT'S BEEN A BUSY YEAR. Last year I hadn't been back from England long when four of us took off on a bit of an overseas trip that took in, among other things, a safari in Tanzania. I ended up in a Maasai village talking, through an interpreter, to those marvellous, tall women in their wonderful blue robes about textiles and the way we make woollens in New Zealand. I gave them a bit of a knitting demonstration. This year, after being in New York in January, we went to Gallipoli in April, which was a moving experience. Like most New Zealand families, we have ancestors who served there, and in France. One of Robert Adam Forrester's sons, Matt, was in the thick of some of the worst of the fighting at Gallipoli, and then on the Western Front in France. He was one of the lucky ones who made it through, but you wonder what it did to him. My own relations Stanley Miles and William Price were both killed in action in the First World War.

In May, I also went on to Paris to attend the Eighth World Congress on Coloured Sheep, as I had been invited to speak about my experiences in bringing naturally coloured yarn to market. I called my paper 'La Laine à l'Haute Coûture — From

Fleece to Fashion', and I decided I would deliver it in French. I had it translated and wrote it out phonetically, and, even though I was no kind of French scholar at all at school, I read it and read it until I could more or less pronounce the words properly. It seemed to go off okay. I didn't see any of the French speakers in the audience wince too much or too obviously, and they maybe seemed genuinely pleased that I had made the effort. And just before I left for Paris on that excursion, Julie passed on an invitation from the US organisers of an international fashion parade that selects designers from around the world and shows them off at some exotic location or another — it's been at the top of the Eiffel Tower and the Grand Canyon. This year it was at the top of the World Trade Centre — and it's covered by CNN, the BBC and about 12 other TV media networks each time.

Unfortunately, I had to decline — there just wouldn't have been time to get everything organised by the time I got back from Paris — but I did ask them to consider me again. Who knows, maybe I'll get to show off my garments at the Taj Mahal or somewhere like that! And meanwhile, I've had a photographer and a journalist here doing a profile for *Life and Leisure* magazine, and Helen and I have each (separately) featured on a SKY TV show, *High Heels in the High Country*. Phew! And then Rural Women New Zealand sought this book.

Now I have time for a bit of a breather, after lambing and before my busy October. I'm pretty tired. A big part of my job as team leader of the BlackHills venture is to reassure my team that they're doing a good job. I have a Christmas function every year, and I invite everyone, from the bloke who shoots rabbits around the place to my knitters, and everyone in between, whether they're the accountant, the shearers, the computer whizz or the vet. But the job of keeping everyone happy is ongoing. I'll tell Russell how great he is every time

I take a thermos of coffee and a couple of scones across the farm. Diane doesn't mind hearing how vital she is to the operation, and I make a point of letting Jude know how good a job she's doing, too. It's all true: I couldn't do what I'm doing without them. But it's exhausting, and in my position I do sometimes wish that there was someone telling me that I'm doing a great job, too.

One nice thing is that I have a significant other in my life again. About this time four years ago, our old mate Stewart Johns came up to have a look at the fence between the house and the home paddock. I dropped off meals-on-wheels and cooked him the odd dinner in return. I noticed as he drove off that he was paying pretty close attention to the state of my other fences. And sure enough, he's strained a bit of wire since then, I've cooked a few dinners and we've pretty much mated up.

There's general approval around here. Stewart lost Diane eight years ago (Jim had been best man at their wedding), and, what with me being on my own after losing Jim, everyone seems pleased for us. Stewart has two lovely daughters, Nicola and Melanie, and grandchildren, and it's been wonderful their allowing me to become part of their lives — a real bonus to me at this stage in my life. There are actually other practical benefits to the arrangement besides the fence maintenance, too, I must admit. All my hoggets now go down to Stewart's place just on the seaward side of Amberley, because, even though he's only down the road from me, the grass grows up to his ankles there when it's still waking up from winter here. The sheep enjoy the cool easterly breeze and a sea view.

Funny how things work out.

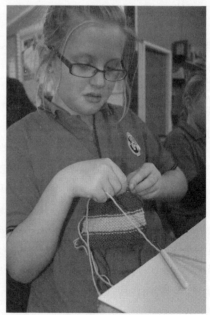

Above Stewart's grandsons, Ryan (5) and Matthew (7). Even budding All Blacks want to knit!

Right Adele Gardner, the daughter of Jolene Gardner who does the BlackHills graphics. Another keen knitter in the making.

A LOT OF GOOD THINGS have come my way through my business. And I try to reciprocate as much as possible, because I reckon that's the way it works. I have polytech students come and do an internship with me, and I just did a fashion parade for a charity group the other day. Since my business has been going well, I've got into sponsoring things a bit. One day on my way home I saw a bunch of cyclists stopped by the road. I recognised one of the people supporting them — an old workmate of mine. I stopped to have a yarn with her and her son, Michael Vink, who was doing pretty well in junior road-cycling competitions. I sponsored him then and there, and he's gone from strength to strength. He was the New Zealand Under-19 champion, and he's a professional rider overseas now. I also sponsor Tryphena Carter, who is a mate of Sarah Heddell, my god-daughter. Tryphena has won a ploughing championship, and what I like about her is that as a young woman she is a stand-out figure in a male-dominated sport. I do things here and there, too, like sponsoring a jump in the showjumping ring, the dog trials, equipment for the North Canterbury disabled, and the Conservation Trust Kiwi egg-recovery programme.

I'm always happy to help people who are labouring under some sort of disadvantage if I can. I've had young men up here who are the children of alcoholics, just helping out around the farm and receiving a little bit of tough love from time to time, which most of them seem to need. Funny thing, last year when we had the big dump of snow, one of those boys rang me up. I hadn't heard from him for well over a year, but he offered to come up and help.

'That'd be marvellous,' I said. 'Do you have a car?'

'Yep,' he said.

Knowing the way these boys think, I thought to ask: 'Is it registered?'

'Yeah, nah. It's not.'

'OK . . . Do you have your licence?'

'Nah. No, I don't.'

My reply was: 'I think you'd better stay put, then. You're a danger on the road — it's snowing.'

Some people need all the help they can get, and at times it's nice to be in a position to give it.

I'VE TOLD THE STORY OF Black Hills as best I can, but of course the story doesn't finish here, and it won't be finished when I'm gone, either. It will be interesting to see what happens with farming in the district, and in New Zealand more generally. If you go to the top of Mount Alexander across the way there, you can literally see the fashions changing. There are the vineyards off towards the coast; they were going to try that across the way from me on a bit of what used to be Karaka, too. Wine stopped being the golden goose that it was 10 years ago. That block's for sale now.

And further off, towards the Hurunui River, you can see where the pines of Balmoral Forest are being cleared to make way for dairy conversion. Pines were the big thing for a while, but the bottom's gone from that market, too. Dairy's 'it', right now, although prices are coming back down after the big payouts they were having, and you do wonder where they'll get the water from over there on those stony river flats. There's a big scheme afoot to store water and dam the Hurunui for irrigation, but it's a sore point with a lot of people, and I don't know whether it will go ahead any time soon.

Meanwhile, those of us running sheep just carry on. The efforts of wool advocates like Michael Mellon and the Wool Advancement Group have yet to pay off, and world wool prices are still pretty depressed. Same with meat. But my little corner of the market just gets better and better.

It's probably a reaction to all the artificial stuff in our lives, but people really respond to the idea of a product that has been handmade from raw materials raised naturally the artisan way, not the industrial way. It's all about traceability, point of origin, and quality assurance. There's a sense that the harder we have tried to be more productive, to make more and sell more, the more we have lost our connection with the stuff that's really important in our lives. That's what BlackHills gives people, I suppose. I'm proud to tell people that my family and the family I married into have a long history of five or six generations on the land. I do things in much the same way as those canny old Scots shepherds used to do things on this land before the chemicals and the machines came along. It's better for the land, it's better for the animals, and it's better for the people working with them. It's no wonder it's better for the consumer, too.

I must go and feed my cat, Thomas, my cattle and my sheepdog, Nugget.

So, there it is. I suppose I must be doing something right. And as long as I'm happy doing it and people are happy that I'm doing it, it's best I just stick to my knitting.

ACKNOWLEDGEMENTS

BEVERLEY

Thank you to Lorraine Hatfull, Maureen Phillips and members of the Rural Women New Zealand Kourawhero branch who nominated me for the Rural Women New Zealand Enterprising Rural Woman Award 2009.

Many thanks to John McCrystal who took up the challenge and conscientiously put the story into words. Sharing the knowledge with him and Doc Sidey helped me gain even more insight into my own district.

Thank you to my sister Noelene for photographs and proof-reading.

Thank you also to Rural Women New Zealand Wellington who asked for the story to be written, and to whom royalties will go. You made this book happen.

JOHN

I have consulted a number of sources and texts in the writing of this story. Early newspapers are a rich repository of information on the being and doings of our forebears: the miraculous technology of Papers Past has made it possible to search this from the comfort of one's own laptop. The Trove online newpaper archive of the National Library of Australia is also a great boon. Te Ara, the online encyclopaedia of New Zealand was useful for fact-checking and biographical detail.

On farming history in general, I consulted Gordon McLauchlan's *The Farming of New Zealand* (1981), and Tom Brooking's *Lands for the People? The Highland Clearances and the Colonisation of New Zealand* (1996). For information on coloured sheep, the proceedings of the 6th World Congress, edited by Roger Lundie and Elspeth Wilkinson and entitled *The World of Coloured Sheep* (2004), was invaluable. On the Corriedale breed of sheep, Tom Burrows' *The Corriedale: A history of the breed and its development in New Zealand* was useful.

On the Forrester family, I relied upon the booklet 'The Descendants of Robert and Louisa Forrester', compiled for a reunion of those descendants held at Hawarden on 3 January 1994. The repository of all knowledge on the Price, Heales and Miles families and their forebears is Beverley's sister, Noelene Quedley, and I am indebted to her for the material she provided, the constant stream of additional information and the fact-checking she undertook. For the history of the Waikari, Hurunui and Amuri districts, I consulted Douglas Cresswell's *Canterbury Tales* (1951), *Squatter and Settler in the Waipara Country* (1952) and *The Story of Cheviot* (1951); Shona McRae's *Hurunui: Source to Sea* (1991); Sheila Crawford's *Sheep and Sheepmen of Canterbury*, 1850-1914 (1949), L G D Acland's *The Early Canterbury Runs* (1952); W J Gardner's *The Amuri: a county history* (1983); and D N Hawkins's, *Beyond* the *Waimakariri: A Regional History* (1957). Almost half of the pioneer history of the Hurunui would have been lost had it not been for the efforts of Doc Sidey in compiling *Our Apron Strings: Women of the Waikari/Hawarden District* (2009).

The history of Glenmark has been well chronicled by David Gee and Herbert Farrant in *Moore of Glenmark* (2006), and Beverley and I were fortunate to spend some time in the company of John McCaskey of Waipara, who has a remarkable collection of photographs of Glenmark in its glory days. A collection of artefacts from New Zealand's forgotten empire are on display in the Hawarden Museum.

I have directly quoted from the following sources: p. 59 Acland (1952) p. 267; p. 62 Crawford (1949) pp. 33-5; p. 65 *Christchurch Press* 30/5/14 p. 2; p. 70 Cresswell (1952) p. 103; pp. 71-2 *Christchurch Press* 30/5/14 p. 2; p. 73 Cresswell (1952) p. 103; p. 74 *Christchurch Press* 30/5/14 p. 2; p. 75 Cresswell (1952) p. 102; p. 79 *Christchurch Press* 9/08/1890 p. 8; p. 82 Cresswell (1952) pp. 103-6; p. 83 *Christchurch Press* 30/5/14 p. 2; p. 83, Cresswell (1952) p. 104; p. 96 http://www.teara.govt.nz/en/biographies/1o2/oconnell-sarah; p. 108 Burrows, p. 6; p. 112 Burrows, p. 8; p. 113 Forrester family history.

In the researching and writing of this book, I have enjoyed the generosity, hospitality and company of a number of people. Doc and Jan Sidey put me up, and the evening I spent discovering that Doc's knowledge of Hurunui history is matched only by his knowledge of cricket was a very pleasant one indeed. Russell and May Hassall and Diane Nurse plied me with tea, biscuits and useful information, for which I am grateful. I enjoyed my stay at the Hurunui Hotel, and of course, I spent many hours in Beverley Forrester's company. Thanks, as ever, to the team at Random House New Zealand — Nicola, Margaret, Leonie and Kate. Last, but not least, thanks to Beverley for her enthusiasm and endless patience.